Practice Papers for SQA Exams

Standard Grade | Credit

Mathematics

ISBN 978-1-84372-773-6

Published by
Leckie & Leckie Ltd, 3rd floor, 4 Queen Street, Edinburgh, EH2 1JE
Tel: 0131 220 6831 Fax: 0131 225 9987
enquiries@leckieandleckie.co.uk www.leckieandleckie.co.uk

A CIP Catalogue record for this book is available from the British Library.

Leckie & Leckie Ltd is a division of Huveaux plc.

Questions and answers in this book do not emanate from SQA. All of our entirely new and original Practice Papers have been written by experienced authors working directly for the publisher.

Introduction

Layout of the Book

This book contains practice exam papers, which mirror the actual SQA exam as much as possible. The layout, paper colour and question level are all similar to the actual exam that you will sit, so that you are familiar with what the exam paper will look like.

The solutions section is at the back of the book. The full worked solution is given to each question so that you can see how the right answer has been arrived at. The solutions are accompanied by a commentary which includes further explanations and advice. There is also an indication of how the marks are allocated and, where relevant, what the examiners will be looking for. Reference is made at times to the relevant sections in Leckie & Leckie's book 'Maths: Credit Level Course Notes'

Revision advice is provided in this introductory section of the book, so please read on!

How To Use This Book

The Practice Papers can be used in two main ways:

1. You can complete an entire practice paper as preparation for the final exam. If you would like to use the book in this way, you can either complete the practice paper under exam style conditions by setting yourself a time for each paper and answering it as well as possible without using any references or notes. Alternatively, you can answer the practice paper questions as a revision exercise, using your notes to produce a model answer. Your teacher may mark these for you.

2. You can use the Topic Index at the front of this book to find all the questions within the book that deal with a specific topic. This allows you to focus specifically on areas that you particularly want to revise or, if you are mid-way through your course, it lets you practise answering exam-style questions for just those topics that you have studied.

Revision Advice

Work out a revision timetable for each week's work in advance – remember to cover all of your subjects and to leave time for homework and breaks. For example:

Day	6pm–6.45pm	7pm–8pm	8.15pm–9pm	9.15pm–10pm
Monday	Homework	Homework	English revision	Chemistry Revision
Tuesday	Maths Revision	Physics revision	Homework	Free
Wednesday	Geography Revision	Modern Studies Revision	English Revision	French Revision
Thursday	Homework	Maths Revision	Chemistry Revision	Free
Friday	Geography Revision	French Revision	Free	Free
Saturday	Free	Free	Free	Free
Sunday	Modern Studies Revision	Maths Revision	Modern Studies	Homework

Make sure that you have at least one evening free a week to relax, socialise and re-charge your batteries. It also gives your brain a chance to process the information that you have been feeding it all week.

Arrange your study time into one hour or 30 minutes sessions, with a break between sessions e.g. 6pm –7pm, 7.15pm –7.45pm, 8pm–9pm. Try to start studying as early as possible in the evening when your brain is still alert and be aware that the longer you put off starting, the harder it will be to start!

Study a different subject in each session, except for the day before an exam.

Do something different during your breaks between study sessions – have a cup of tea, or listen to some music. Don't let your 15 minutes expanded into 20 or 25 minutes though!

Have your class notes and any textbooks available for your revision to hand as well as plenty of blank paper, a pen, etc. You should take note of any topic area that you are having particular difficulty with, as and when the difficulty arises. Revisit that question later having revised that topic area by attempting some further questions from the exercises in your textbook.

Revising for a Maths Exam is different from revising for some of your other subjects. Revision is only effective if you are trying to solve problems. You may like to make a list of 'Key Questions' with the dates of your various attempts (successful or not!). These should be questions that you have had real difficulty with.

Key Question	1st Attempt		2nd Attempt		3rd Attempt	
Textbook P56 Q3a	18/2/10	×	21/2/10	√	28/2/10	√
Practice Exam A Paper 1 Q5	25/2/10	×	28/2/10	×	3/3/10	
2008 SQA Paper, Paper 2 Q4c	27/2/10	×	2/3/10			

The method for working this list is as follows:

1. Any attempt at a question should be dated.

2. A tick or cross should be entered to mark the success or failure of each attempt.

3. A date for your next attempt at that question should be entered:
   for an unsuccessful attempt – 3 days later
   for a successful attempt – 1 week later

4. After two successful attempts remove that question from the list (you can assume the question has been learnt!)

Using 'The List' method for revising for your Maths Exam ensures that your revision is focused on the difficulties you have had and that you are actively trying to overcome these difficulties.

Finally forget or ignore all or some of the advice in this section if you are happy with your present way of studying. Everyone revises differently, so find a way that works for you!

Transfer Your Knowledge

As well as using your class notes and textbooks to revise, these practice papers will also be a useful revision tool as they will help you to get used to answering exam style questions. You may find as you work through the questions that you find an example that you haven't come across before. Don't worry! There may be several reasons for this. You may have come across a question on a topic that you have not yet covered in class. Check with your teacher to find out if this is the case. Or it may be the case that the wording or the context of the question is unfamiliar. This is often the case with reasoning questions in the Maths Exam. Once you have familiarised yourself with the worked solutions, in most cases you will find that the question is using mathematical techniques with which you are familiar. In either case you should revisit that question later to check that you can successfully solve it.

Trigger Words

In the practice papers and in the exam itself, a number of 'trigger words' will be used in the questions. These trigger words should help you identify a process or a technique that is expected in your solution to that part of the question. If you familiarise yourself with these trigger words, it will help you to structure your solutions more effectively.

Trigger Words	Meaning/Explanation
Evaluate	Carry out a calculation to give an answer that is a value.
Hence	You must use the result of the previous part of the question to complete your solution. No marks will be given if you use an alternative method that does not use the previous answer.

Simplify	This means different things in different contexts:
	   Surds: reduce the number under the root sign to the smallest possible by removing square factors.
	   Fractions: one fraction, cancelled down, is expected.
	   Algebraic expressions: get rid of brackets and gather all like terms together.

Give your answer to ...	This is an instruction for the accuracy of your final answer. These instructions must be followed or you will lose a mark.
Algebraically	The method you use must involve algebra e.g. you must solve an equation or simplify an algebraic equation. It is usually stated to avoid trial-and-improvement methods or reading answers from your calculator.
Justify your answer	This is a request for you to indicate clearly your reasoning. Will the examiner know how your answer was obtained?
Show all your working	Marks will be allocated for the individual steps in your working. Steps missed out may lose you marks.

In the Exam

Watch your time and pace yourself carefully. Some questions you will find harder than others. Try not to get stuck on one question as you may later run out of time. Rather return to a difficult question later. Remember also that if you have spare time towards the end of your exam, use this time to check through your solutions. Often mistakes are discovered in this checking process and can be corrected.

Become familiar with the exam instructions. The practice papers in this book have the exam instructions at the front of each exam. Also remember that there is a formuae list to consult. You will find this at the front of your exam paper. However, even though these formulae are given to you, it is important that you learn them so that they are familiar to you. If you are continuing with Mathematics next session it will be assumed that these formulae are known in next year's exam!

Read the question thoroughly before you begin to answer it – make sure you know exactly what the question is asking you to do. If the question is in sections e.g. 15a, 15b, 15c, etc, then it is often the case that answers obtained in the earlier sections are used in the later sections of that question.

When you have completed your solution read it over again. Is your reasoning clear? Will the examiner understand how you arrived at your answer? If in doubt then fill in more details.

If you change your mind or think that your solution is wrong, don't score it out unless you have another solution to replace it with. Solutions that are not correct can often gain some of the marks available. Do not miss working out. Showing step-by-step working will help you gain maximum marks even if there is a mistake in the working.

Use these resources constructively by reworking questions later that you found difficult or impossible first time round. Remember: success in a Maths exam will only come from actively trying to solve lots of questions and only consulting notes when you are stuck. Reading notes alone is not a good way to revise for your Maths exam. Always be active, always solve problems.

Good luck!

Topic Index

Topic	A Paper 1	A Paper 2	B Paper 1	B Paper 2	C Paper 1	C Paper 2
Number & Money						
• Decimals	Q1		Q1		Q1	
• Fractions	Q2		Q3		Q2	
• Percentages	Q6	Q4		Q1		Q2
• Ratios			Q10			Q7
• Scientific Notation		Q1				Q1

Topic	A Paper 1	A Paper 2	B Paper 1	B Paper 2	C Paper 1	C Paper 2
Shape & Measure						
• Bearings			Q3			Q4
• Circles	Q12	Q8		Q10		Q5, Q6
• Pythagoras' Theorem		Q8		Q10		Q6a, Q12
• Similarity				Q9		Q12
• Solids	Q9	Q13	Q4		Q5	
Trigonometry						
• Area Formula		Q9		Q7		
• Cosine Rule		Q7			Q7	
• Right-angled Triangles		Q6		Q6		
• Sine Rule		Q5		Q3		Q4
• Trig Equations		Q10				Q8
• Trig Graphs	Q9					
Algebraic Relationships						
• Algebraic Fractions	Q4				Q5	
• Brackets			Q4			Q11c
• Factorisation			Q2		Q5a	
• Formulae			Q13		Q6	Q9
• Function Notation			Q5		Q3	
• Indices	Q11a		Q12a		Q11	
• Inequalities					Q4	
• Linear Equations			Q5	Q11	Q13	Q9
• Quadratic Equations		Q11			Q9	Q3
• Sequences/ *n*th terms			Q11		Q5	Q11
• Simultaneous Equations	Q8		Q7			
• Substitution	Q3				Q3	
• Surds	Q11b		Q12b		Q10	
• Variation						Q10
Graphical Relationships						
• Linear Graphs	Q10	Q2	Q6		Q12	
• Quadratic Graphs				Q8		
Statistics & Probability						
• Boxplots	Q5		Q9		Q8	
• Probability	Q7		Q8			
• Stand. Dev. & Quartiles		Q3		Q2	Q8	

In the answers, there are references to the pages of Leckie & Leckie's *Credit Maths Revision Notes* (ISBN 978-1-84372-079-9). These will help you learn more about any topics you might find difficult. You could also check out *Questions in Credit Maths* (ISBN 978-1-84372-147-5) for graded practice of these topics.

Practice Exam A

Mathematics | Standard Grade | Credit

Practice Papers
For SQA Exams

**Exam A
Credit Level
Paper 1
Non-calculator**

You are allowed 55 minutes to complete this paper

Do **not** use a calculator.

Try to answer all of the questions in the time allowed, including all of your working.

Full marks will only be awarded where your answer includes any relevant working.

Scotland's leading educational publishers

FORMULAE LIST

Standard Deviation: $s = \sqrt{\dfrac{\Sigma(x - \overline{x})^2}{n-1}} = \sqrt{\dfrac{\Sigma x^2 - (\Sigma x)^2/n}{n-1}}$, where n is the sample size.

Area of a triangle: $\text{Area} = \dfrac{1}{2}\,ab\sin C$

Sine Rule: $\dfrac{a}{\sin A} = \dfrac{b}{\sin B} = \dfrac{c}{\sin C}$

Cosine Rule: $a^2 = b^2 + c^2 - 2bc\cos A$ or $\cos A = \dfrac{b^2 + c^2 - a^2}{2bc}$

The roots of $ax^2 + bx + c = 0$ are $x = \dfrac{-b \pm \sqrt{(b^2 - 4ac)}}{2a}$

1. Evaluate

$$5{\cdot}1 \div (8{\cdot}21 - 5{\cdot}21)$$

	KU	RE
	2	

2. Evaluate

$$2\tfrac{2}{3} - 1\tfrac{1}{5} \times \tfrac{1}{3}$$

3

3. Q = R − S where R = $3a^2$ and S = b^2
Calculate the value of Q when $a = -2$ and $b = 3$

2

4. Express as a single fraction in its simplest form

$$\frac{7}{x} - \frac{3}{x+1}$$

3

5. The number of cases of flu treated at 14 health centres on the 1st November was recorded.

$$\begin{array}{ccccccc} 2 & 2 & 5 & 12 & 13 & 26 & 27 \\ 23 & 20 & 19 & 11 & 8 & 3 & 6 \end{array}$$

Draw a suitable statistical diagram to illustrate the median and the quartiles of this data.

4

KU	RE

6. A retailer is offering a 20% discount on the normal price for all 21-inch computer monitors.

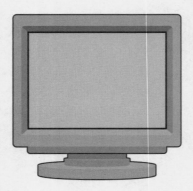

This is a 21-inch model and is now being sold at a reduced price of £680.

What is the normal price of this monitor?

3

7. This dice is a regular dodecahedron. Its 12 faces are numbered 1 to 12.

This dice is a regular octahedron. Its 8 faces are numbered 1 to 8.

On each dice each number has an equal chance of being rolled. Haleem has to roll a multiple of 3 to win the game he is playing. Which dice would give him a better chance of winning? Show clearly all your working.

3

8. Number Triangles

The rule in a Number Triangle is A = B + C

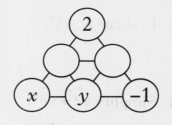

	KU	RE

(a) Use this rule to copy and complete this Number Triangle.

Show that $x + 2y = 3$

2

(b) The same Number Triangle rule is used for this Number Triangle. Write down another equation with x and y.

2

(c) In both these Number Triangles, x and y have the same values. Find the values of x and y.

3

9. The graph of $y = p \sin qx°$, $0 \le x \le 120$ is shown below.

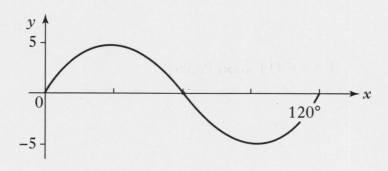

Write down the values of p and q.

2

	KU	RE

	KU	RE

10. Two variables x and y are connected by the relationship $y = px + q$. It is known that $p > 0$ and $q < 0$.

Sketch a possible graph of y against x to illustrate this relationship.

RE: 3

11. (*a*) Evaluate $8^{\frac{2}{3}} - 8^0$

KU: 2

(*b*) Simplify $\sqrt{2} + \sqrt{8}$

KU: 2

12. The circular clock face on Big Ben is set in a square frame as shown in the diagram.

The tip of the minute hand travels a distance of 27 metres every hour.

Show that the perimeter of the square frame is exactly $\dfrac{108}{\pi}$ metres.

RE: 4

[End of Question Paper]

Mathematics | Standard Grade | Credit

Practice Papers
For SQA Exams

Exam A
Credit Level
Paper 2

You are allowed 1 hour, 20 minutes to complete this paper

A calculator can be used.

Try to answer all of the questions in the time allowed, including all of your working.

Full marks will only be awarded where your answer includes any relevant working.

Scotland's leading educational publishers

FORMULAE LIST

Standard Deviation: $s = \sqrt{\dfrac{\Sigma(x-\bar{x})^2}{n-1}} = \sqrt{\dfrac{\Sigma x^2 - (\Sigma x)^2/n}{n-1}}$, where n is the sample size.

Area of a triangle: $\text{Area} = \dfrac{1}{2}\,ab\sin C$

Sine Rule: $\dfrac{a}{\sin A} = \dfrac{b}{\sin B} = \dfrac{c}{\sin C}$

Cosine Rule: $a^2 = b^2 + c^2 - 2bc\cos A$ or $\cos A = \dfrac{b^2 + c^2 - a^2}{2bc}$

The roots of $ax^2 + bx + c = 0$ are $x = \dfrac{-b \pm \sqrt{(b^2 - 4ac)}}{2a}$

1. On August 27, 2003 Mars was 5.6×10^7 km from Earth, its closest approach for 60 000 years.

How long would it take to drive this distance in a car (if this were possible!) at an average speed of 120 kilometres per hour?

Give your answer in days in scientific notation.

KU **4**

2. To set up a live video link for a conference a University is charged an initial connection set-up fee of £120. Thereafter they are charged at a fixed rate per hour while the conference lasts. They were charged £245 for the 5 hour conference.

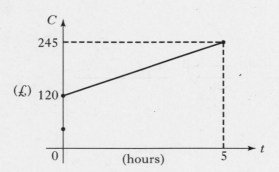

The above graph represents the cost (£C) against the time (t hours).

(a) Find the equation of the line in terms of C and t.

RE **3**

(b) The next day the University was charged £282·50 for a similar video conference. How long did this conference last?

RE **3**

3. A group of eight gap-year students in Malawi were badly bitten by mosquitoes and contracted Malaria. Normally fever starts after 10 days.

The time, in days, for the onset of fever in this group were:

7, 10, 8, 9, 13, 7, 10, 8

Calculate the mean and standard deviation of these times.

KU **4**

	KU	RE

4. In 2001 the Russian Space Station Mir was destroyed as it burned up in the upper atmosphere.

It had been losing altitude by 6% every month.

At the start of December 2000 it's altitude was 340 km.

What was its altitude by the start of March 2001?

(KU: 3)

5. On a snooker table the distance between two pockets A and B is 1·78 metres.

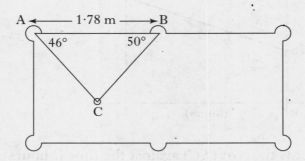

From pocket A the 'line-of-sight' to the ball at C makes an angle of 46° with the edge cushion.

Similarly the 'line-of-sight' from pocket B to the ball at C makes an angle of 50° with the edge cushion as shown.

Calculate the distance of the ball at C from pocket A.

(KU: 4)

6. The diagram shows a triangular flag with a shaded triangular design.

The flag is in the shape of an isosceles triangle ACD.

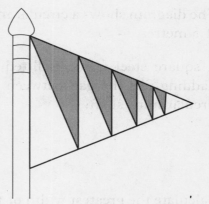

In the design the largest side of each shaded triangle is at right angles to the lower edge of the flag.

Side AB is 80 cm in length, as shown in the diagram. The tip of the flag forms a 45° angle.

Calculate the length AC of the flag at the flagpole.

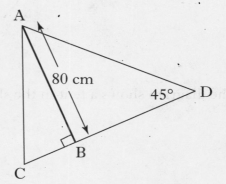

7. The diagram shows a gear-slider mechanism.

Rod AB = 10·5 cm

Rod BC = 18 cm

angle ABC = 95°

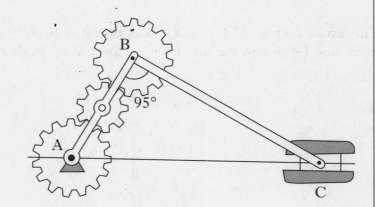

Calculate the length of the slide AC to 1 decimal place.

KU	RE
	5
	4

8. The diagram shows a circular mine shaft with a radius of 5 metres.

A square steel frame is fitted for a lift. A concrete cladding fills the gap between the lift frame and the circular mine shaft.

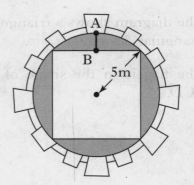

Calculate the greatest width of the concrete cladding (AB in the diagram) giving your answer to 3 significant figures.

9. The diagram shows a tent in the shape of a triangular prism.

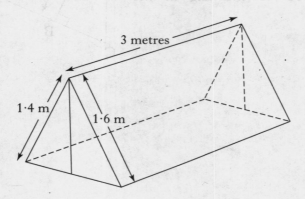

This cross-section of the tent shows that it is slightly tilted to the left. The two sides, one 1·4 m and the other 1·6 m, meet at an angle of 50° at the top.

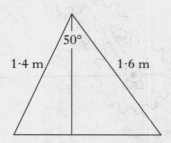

Calculate the volume of the tent.

KU	RE

10. Solve algebraically the equation

$$3 \cos x° + 4 = 3 \qquad 0 \le x < 360$$

KU **3**

11. The diagram shows a swimming pool of width x metres surrounded by a concrete path.

The length of the swimming pool is 2 metres more than its width.

The surrounding concrete path is 1 metre wide along the length and 2 metres wide along the width as shown in the diagram.

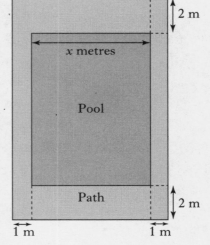

The area of the pool is the same as the area of the concrete path.

(a) Show that $x^2 - 4x - 12 = 0$

RE **4**

(b) Hence find the dimensions of the pool.

RE **3**

[End of Question Paper]

KU	RE

Practice Exam B

Mathematics | Standard Grade | Credit

Practice Papers
For SQA Exams

Exam B
Credit Level
Paper 1
Non-calculator

You are allowed 55 minutes to complete this paper

Do **not** use a calculator.

Try to answer all of the questions in the time allowed, including all of your working.

Full marks will only be awarded where your answer includes any relevant working.

Scotland's leading educational publishers

FORMULAE LIST

Standard Deviation: $s = \sqrt{\dfrac{\Sigma(x - \overline{x})^2}{n-1}} = \sqrt{\dfrac{\Sigma x^2 - (\Sigma x)^2/n}{n-1}}$, where n is the sample size.

Area of a triangle: $\text{Area} = \dfrac{1}{2} ab \sin C$

Sine Rule: $\dfrac{a}{\sin A} = \dfrac{b}{\sin B} = \dfrac{c}{\sin C}$

Cosine Rule: $a^2 = b^2 + c^2 - 2bc \cos A$ or $\cos A = \dfrac{b^2 + c^2 - a^2}{2bc}$

The roots of $ax^2 + bx + c = 0$ are $x = \dfrac{-b \pm \sqrt{(b^2 - 4ac)}}{2a}$

1. Evaluate

$$8{\cdot}82 - 2{\cdot}9 \times 3$$

KU	RE
2	

2. Factorise

$$2x^2 + 3x - 2$$

2	

3. Evaluate

$$\frac{2}{3}\left(1\tfrac{1}{2} - \tfrac{1}{3}\right)$$

2	

4. Simplify

$$x(2x - 1) - x(2 - 3x)$$

3	

5. $f(x) = \dfrac{6}{x}$, $x \neq 0$

(a) Evaluate $f(-3)$

1	

(b) Given that $f(a) = 3$, find a.

2	

6. In this diagram the line AB has gradient $-\frac{1}{2}$ and it cuts the y-axis at the point $(0, 3)$.

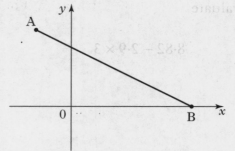

(a) Write down the equation of the line AB.

(b) Point A has coordinates $(-2, a)$.

Find the value of a.

7. Protons and Neutrons are subatomic particles. Inside each atom there is a nucleus made up from Protons and Neutrons. These subatomic particles themselves are made up from two types of quarks: 'up' quarks and 'down' quarks. The electrical charge of a proton or neutron is the sum of the electrical charges of its quarks.

PROTON

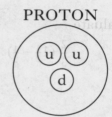

(a) A Proton is made from two 'up' quarks and one 'down' quark and has a total electrical charge of 1 unit.

Write down an algebraic equation to illustrate this.

NEUTRON

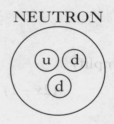

(b) A Neutron is made from one 'up' quark and two 'down' quarks and has a total electrical charge of zero.

Write down an algebraic equation to illustrate this.

(c) Find the electrical charge of an 'up' quark. Show clearly your reasoning.

KU	RE
2	
	2
1	
1	
	3

	KU	RE

8. Mia is in class 3M1 for Maths. There are three S3 Maths sets. The table shows the numbers of boys and girls in each.

	3M1	3M2	3M3
Boys	12	10	14
Girls	16	14	6

(a) A pupil is chosen at random from Mia's class. What is the probability that a boy was chosen?

1

(b) At the end of term the three classes were put together in the school hall to watch a Maths film. If a pupil is chosen at random from the audience what is the probability now that a boy was chosen?

1

9. A random sample of cars on Scotland's roads were tested for fuel efficiency. The measure used was average miles per gallon or mpg. The results are shown in this boxplot:

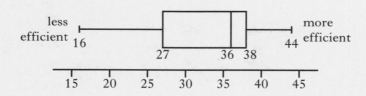

What percentage of sampled cars were less efficient than 38 mpg?

1

10. 'Clearlawn' mosskiller/fertiliser treatment contains nitrogen, phosphorus and potassium mixed in the ratio 3:2:1.

(a) Mr Clevedon's lawn requires 12 kg of nitrogen. Using 'Clearlawn' treatment how much potassium would his lawn get?

1

(b) He decides to use 'Clearlawn' and finds the treatment is sold in 7 kg packets. How many should he buy to treat his lawn? Show clearly your reasoning.

3

	KU	RE

11. A sequence of numbers has the following first few terms:

$$t_1 = 1$$
$$t_2 = 7$$
$$t_3 = 19$$
$$t_4 = 37$$

(a) Complete the calculation of S_3 and S_4 in this pattern:

$$S_1 = t_1 \qquad\qquad = 1 \qquad\qquad = 1$$
$$S_2 = t_1 + t_2 \qquad\qquad = 1 + 7 \qquad\qquad = 8$$
$$S_3 = t_1 + t_2 + t_3 \qquad = \ldots\ldots\ldots \qquad = \ldots$$
$$S_4 = \ldots\ldots\ldots\ldots \qquad = \ldots\ldots\ldots\ldots \qquad = \ldots$$

                                                    2

(b) Suggest a formula for S_n in terms of n **1**

(c) Hence find a formula for t_{n+1} the '$n + 1$st' term of the sequence

12. (a) Evaluate $9^{-\frac{1}{2}}$ **2**

(b) Simplify $\sqrt{t} \times t^2$ **2**

13. The diagram shows a ball-sorting device. Over-sized or under-sized balls are detected by a sensor which activates the trap door to remove them.

The proper-sized balls are collected in the collecting trough.

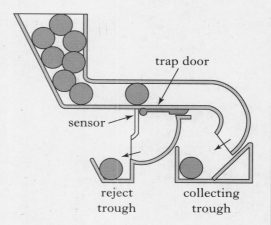

The collecting trough is in the shape of a cuboid with a diagonal shute-plate as shown in the diagram below with all measurements in centimetres.

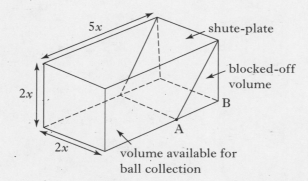

One quarter of the volume of the cuboid is unavailable for collecting as it is blocked off by the shute-plate.

Using the dimensions as shown in the diagram find, in terms of x, the length AB.

4

[End of Question Paper]

Mathematics | Standard Grade | Credit

Practice Papers
For SQA Exams

Exam B
Credit Level
Paper 2

You are allowed 1 hour, 20 minutes to complete this paper

A calculator can be used.

Try to answer all of the questions in the time allowed, including all of your working.

Full marks will only be awarded where your answer includes any relevant working.

Scotland's leading educational publishers

FORMULAE LIST

Standard Deviation: $s = \sqrt{\dfrac{\Sigma(x-\overline{x})^2}{n-1}} = \sqrt{\dfrac{\Sigma x^2 - (\Sigma x)^2/n}{n-1}}$, where n is the sample size.

Area of a triangle: $\text{Area} = \dfrac{1}{2}\,ab\sin C$

Sine Rule: $\dfrac{a}{\sin A} = \dfrac{b}{\sin B} = \dfrac{c}{\sin C}$

Cosine Rule: $a^2 = b^2 + c^2 - 2bc\cos A$ or $\cos A = \dfrac{b^2 + c^2 - a^2}{2bc}$

The roots of $ax^2 + bx + c = 0$ are $x = \dfrac{-b \pm \sqrt{(b^2 - 4ac)}}{2a}$

	KU	RE

1. The total emissions of Greenhouse Gases by the USA in 2007 amounted to the equivalent of 7·2 million tonnes of carbon dioxide. If the annual increase in emissions is 1·2%, calculate the total amount of emissions of Greenhouse Gases by the USA expected in 2010. Give your answer in millions of tonnes to 2 significant figures. **4**

2. The amounts, in £, spent by a sample of six diners at a restaurant one Saturday evening were:

£44, £47, £38, £97, £40, £52

(*a*) Find the mean amount spent. **1**

(*b*) Find the standard deviation of the amounts spent. **2**

(*c*) On a weekday evening at the same restaurant the standard deviation of the amounts spent was £8·40.

Make one valid comparison between the amounts spent by diners on a weekday evening and a Saturday evening. **1**

3. On this map of Fife, Leven lies due West of Elie.

Cupar is 11·9 km from Leven on a bearing of 004°.

Cupar is on a bearing of 320° from Elie.

How far is Elie from Leven? Do not use a scale drawing. **4**

	KU	RE

4. An oil drum is in the shape of a cylinder with diameter 58 cm and height 97 cm.

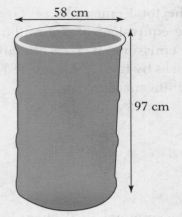

58 cm

97 cm

(*a*) Calculate the volume of the drum giving your answer to the nearest litre.

3

(*b*) If 64 litres are poured out of a full drum what will be the depth of the remaining oil in the drum?

2

5. The 'hex' numbers form a sequence of positive integers. The *n*th 'hex' number is given by $1 - 3n + 3n^2$.
61 is a 'hex' number. Which one is it in the sequence?

4

6. This diagram shows a linkage mechanism in a machine.

The lengths and angles between the various pivot points A, B, C and D are shown below.

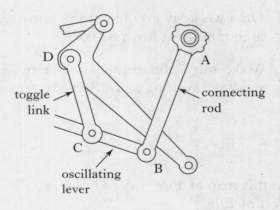

D

toggle link

C

oscillating lever

A

connecting rod

B

AB = 9 cm

Angle CDA = Angle ABC = 90°

Angle BAC = 23°

Angle CAD = 35°

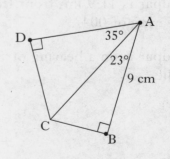

D

A

35°

23°

9 cm

C

B

Calculate the length of the toggle link CD.

4

7. The diagram shows the design for a triangular pendant.

The area of the central triangle PQR is 4·2 cm²

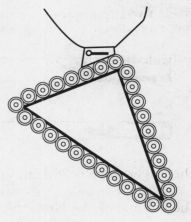

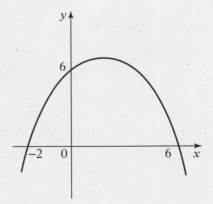

PR = 2·5 cm

PQ = 3·4 cm

Calculate the size of the acute angle QPR at the top of the pendant.   **3**

8. The diagram below shows the graph of a quadratic function with the equation

$$y = k\,(x - p)\,(x - q) \text{ where } p < q$$

The graph cuts the x-axis at the points $(-2, 0)$ and $(6, 0)$ and cuts the y-axis at the point $(0, 6)$.

(*a*) Write down the values of p and q.   **2**

(*b*) Calculate the value of k.   **2**

(*c*) Find the coordinates of the maximum turning point of the function.   **2**

KU	RE

9. These two organ pipes are mathematically similar in shape.

The larger pipe is 240 cm in length and the smaller pipe is 180 cm in length.

The volume of the larger pipe is 43 litres.

Calculate the volume of the smaller pipe to the nearest litre.

3

10. This spanner head is circular with a symmetrical 5-sided gap.

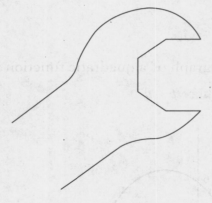

The diagram below shows the dimensions of the circular head. C is the centre of the circle which has diameter 5 cm. The head measures 4·8 cm across as shown.

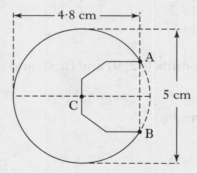

Calculate the width of the gap AB.

4

11. Marcus travelled from St Andrews to Thurso in two stages.

(*a*) In the first stage of his journey he covered 180 miles in x hours.

Find, in terms of x, his average speed.

1

(*b*) He covered the 60 miles of the second stage of his journey in 2 hours less time than the first stage.

Find an expression for his average speed for the second stage of his journey.

1

(*c*) His average speed on both stages of his journey was the same. Calculate the time taken for the whole of his journey.

3

[End of Question Paper]

Practice Exam C

Mathematics | Standard Grade | Credit

Practice Papers
For SQA Exams

Exam C
Credit Level
Paper 1
Non-calculator

You are allowed 55 minutes to complete this paper

Do **not** use a calculator.

Try to answer all of the questions in the time allowed, including all of your working.

Full marks will only be awarded where your answer includes any relevant working.

Scotland's leading educational publishers

FORMULAE LIST

Standard Deviation: $s = \sqrt{\dfrac{\Sigma(x - \bar{x})^2}{n-1}} = \sqrt{\dfrac{\Sigma x^2 - (\Sigma x)^2/n}{n-1}}$, where n is the sample size.

Area of a triangle: $\text{Area} = \dfrac{1}{2}\, ab \sin C$

Sine Rule: $\dfrac{a}{\sin A} = \dfrac{b}{\sin B} = \dfrac{c}{\sin C}$

Cosine Rule: $a^2 = b^2 + c^2 - 2bc \cos A$ or $\cos A = \dfrac{b^2 + c^2 - a^2}{2bc}$

The roots of $ax^2 + bx + c = 0$ are $x = \dfrac{-b \pm \sqrt{(b^2 - 4ac)}}{2a}$

	KU	RE

1. Evaluate

$$1 \cdot 3 \times (4 \cdot 92 + 0 \cdot 08)$$

KU: 2

2. Evaluate

$$\frac{2}{3} \div 1\frac{1}{3}$$

KU: 2

3. Given that $f(x) = x(2 - x)$, evaluate $f(-1)$

KU: 2

4. Solve the inequality $3 + 2x < 4(x + 1)$

RE: 3

5. (*a*) Factorise $9y^2 - 4$

KU: 1

   (*b*) Hence simplify

$$\frac{9y^2 - 4}{15y - 10}$$

RE: 2

6. $P = \dfrac{W}{4A}$

Change the subject of this formula to A

KU: 2

	KU	RE

7. The triangle below shows the distances between Cupar (C), Leven (L) and St Andrews (S).

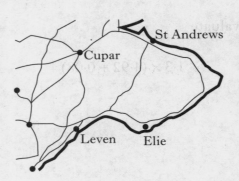

CS = 8 miles

CL = 10 miles

LS = 12 miles

Show that cos C = $\frac{1}{8}$

3

8. A survey was carried out to compare the punctuality of the trains run by two train companies. 10 trains at random were recorded with the 'number of minutes late' being recorded. The results for the two companies were:

Company A					**Company B**				
11	6	1	1	2	4	1	1	3	1
3	9	7	5	5	5	0	8	0	9

Draw an appropriate statistical diagram to compare the two sets of data.

3

9. Part of the graphs $y = f(x)$ and $y = g(x)$ are shown where:

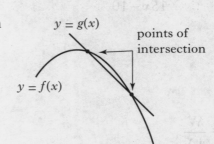

$$f(x) = 4 + 3x - x^2$$

$$g(x) = 10 - 2x$$

Find the x-coordinates of the two points of intersection by solving the equation $f(x) = g(x)$.

3

	KU	RE

10. Simplify fully

$$\sqrt{(\sqrt{18})^2 + (\sqrt{6})^2}$$

KU: 2

11. Express in its simplest form:

$$x^3 \times (x^{-1})^{-2}$$

KU: 2

12. A baby measured 20 inches in length at birth. The baby's length was recorded at regular intervals and a graph plotted.

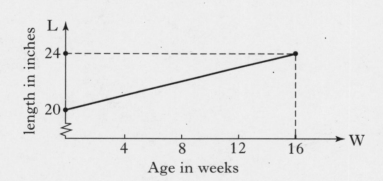

The graph is linear and shows that at 16 weeks the baby's length is 24 inches.

Find the equation of the straight line graph in terms of W and L.

RE: 4

13. Carrie is a tea blender for a large international tea company.

She has recently bought quantities of Kenyan and Rwandan tea and has created two different blends from these teas. They are made in 5 kg packets.

Blend A: 2 kg of Kenyan and 3 kg of Rwandan costing 6·10 euros.

Blend B: 3 kg of Kenyan and 2 kg of Rwandan costing 5·90 euros.

Let the cost of Kenyan be x euros per kg and the cost of Rwandan be y euros per kg.

(a) Write down an algebraic equation to illustrate the make up and cost of Blend A.

(b) Write down a similar equation for Blend B.

(c) She creates a third blend as follows:

Blend C: 4 kg of Kenyan and 1 kg of Rwandan.

Find the cost of 5 kg of Blend C.

KU	RE
1	
1	
	4

[End of Question Paper]

Mathematics | Standard Grade | Credit

Practice Papers
For SQA Exams

Exam C
Credit Level
Paper 2

You are allowed 1 hour, 20 minutes to complete this paper

A calculator can be used.

Try to answer all of the questions in the time allowed, including all of your working.

Full marks will only be awarded where your answer includes any relevant working.

Scotland's leading educational publishers

FORMULAE LIST

Standard Deviation: $s = \sqrt{\dfrac{\Sigma(x - \bar{x})^2}{n-1}} = \sqrt{\dfrac{\Sigma x^2 - (\Sigma x)^2/n}{n-1}}$, where n is the sample size.

Area of a triangle: $\text{Area} = \dfrac{1}{2} ab \sin C$

Sine Rule: $\dfrac{a}{\sin A} = \dfrac{b}{\sin B} = \dfrac{c}{\sin C}$

Cosine Rule: $a^2 = b^2 + c^2 - 2bc \cos A$ or $\cos A = \dfrac{b^2 + c^2 - a^2}{2bc}$

The roots of $ax^2 + bx + c = 0$ are $x = \dfrac{-b \pm \sqrt{(b^2 - 4ac)}}{2a}$

1. In 2009 scientists at Stanford University broke the record for the smallest writing. They wrote 'SU'. Each letter had width 3×10^{-8} cm.

   In letters this size the entire Bible could be written in a line $1 \cdot 05 \times 10^{-1}$ cm long.

   Calculate the number of letters in the Bible. Give your answer in scientific notation.

   2

2. Mr Middleton was informed by 'Petcare' that his annual premium for his Pet insurance had been increased by $6\frac{1}{2}\%$.

   His annual premium was now £87·82

   Calculate his annual premium before this increase.

   3

3. Solve the equation

   $$3x^2 - x - 5 = 0$$

   Give your answers correct to 1 decimal place.

   4

KU | RE

4. From Dublin the bearing of Edinburgh is 035° and the bearing of London is 118°

From Edinburgh the bearing of London is 158°

The distance from Dublin to Edinburgh is 343 km

Calculate the distance from Edinburgh to London.

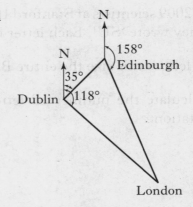

5. A cylindrical plastic water pipe has a uniform thickness of $2\frac{1}{2}$ cm as is shown in the cross-sectional diagram below.

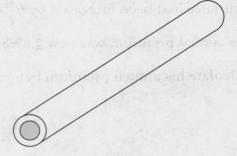

The outside diameter of the pipe is 30 cm and the inside diameter is 25 cm.

(*a*) Calculate the area of plastic in the cross-section.

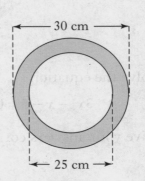

(*b*) 1 cm³ of plastic weighs 0·12 grams.

Calculate the weight of a 1 metre length of pipe.

KU	RE
	5
3	
2	

6. An aircraft fuselage has a circular cross-section of diameter 4·6 metres.

The passenger compartment floor is 1·2 metres above the lowest point of the cargo compartment as shown below:

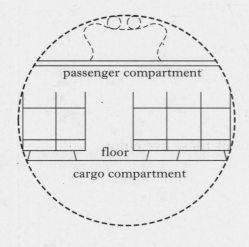

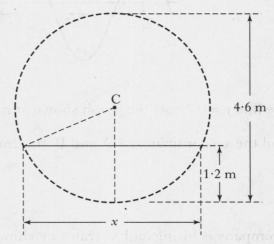

In the diagram above C is the centre of the cross-section.

(a) Calculate x, the width of the passenger compartment floor.

3

(b) The passenger compartment ceiling is the same width as the floor. How high above the floor is it?

1

7. A chocolate syrup manufacturer uses a recipe that requires mixing honey and cocoa in the ratio 9:2 by weight.

The manufacturer sells the syrup in 1 kg jars.

At the end of the week they have $49\frac{1}{2}$ kg of honey and 12 kg of cocoa left.

What is the maximum number of 1 kg jars of syrup that can be made with these remaining ingredients?

3

	KU	RE

8. The diagram shows part of the graph $y = \cos x°$

The line $y = 0.2$ cuts the graph shown at five points.

Find the x-coordinates of A and B, the 2nd and 5th of these five points.

RE: 3

9. A company is hiring an Excavator machine.

They are considering two Plant Hire companies, 'Earthmove' and 'Trenchers'.

The hire rates for these companies are as follows:

'Earthmove'	**'Trenchers'**
Delivery/Collection charge £64	Delivery/Collection charge £28
Hourly rate: £30	Hourly rate: £34

(a) Calculate the cost of a 2 hour hire from each of the companies.

KU: 1

(b) For a 20 hour hire which company is cheaper?

RE: 1

(c) For each company find a formula for the cost of a 'n' hour hire.

RE: 2

(d) 'Trenchers' claim that they are the cheaper of the two companies.

Find algebraically the greatest number of hours of hire for this claim to be true.

RE: 2

10. In comparing stars to the sun, astronomers use this information.

The diameter, D, of the star varies directly as the square root of its luminosity, L, and inversely as the square of its temperature, T.

(a) Write down a formula for D in terms of L and T.   1

(b) If the luminosity is multiplied by 4 and the temperature is doubled, what happens to the diameter?   2

11. Here is a number pattern:

$$2 + 1 = 1 \times 3 - 2 \times 0$$
$$4 + 1 = 2 \times 4 - 3 \times 1$$
$$6 + 1 = 3 \times 5 - 4 \times 2$$

(a) Write down the 4th line of this pattern.   1

(b) Write down the nth line of this pattern.   2

(c) Hence show algebraically that this pattern is always true.   1

KU	RE

12. This diagram shows the supporting wooden structure for a roof.

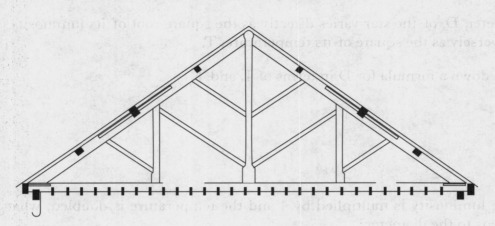

	KU	RE

The diagram on the right shows the dimensions of one of the central sections bounded by the two vertical posts AG and DF.

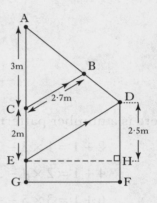

The 'web truss' BC is 2·7 m in length and is parallel to the other 'web truss' DE.

Section AC of the 'King post' AG is 3 m in length and the lower section CE is 2 m in length.

EH is a horizontal line parallel to the 'bottom chord' GF with DH 2·5 m in length.

Calculate the distance, GF, between the two vertical posts.

5

[End of Question Paper]

Worked Answers

Practice Exam A: Paper 1 Worked Answers

Q1 $5 \cdot 1 \div (8 \cdot 21 - 5 \cdot 21)$

$= 5 \cdot 1 \div 3$ ✔

$= 1 \cdot 7$ ✔

2 marks

Order of Operations
- Calculations in brackets should be done first. The subtraction is done before the division in this case

Division of Decimals
- No Calculator is allowed in Paper 1
- Working: $3\overline{)5 \cdot ^2 1}$ gives $1 \cdot 7$

Q2 $2\frac{2}{3} - 1\frac{1}{5} \times \frac{1}{3}$

$= \frac{8}{3} - \left(\frac{6}{5} \times \frac{1}{3}\right)$ ✔

$= \frac{8}{3} - \frac{6}{15}$ ✔

$= \frac{40}{15} - \frac{6}{15}$

$= \frac{40 - 6}{15}$

$= \frac{34}{15}$

$= 2\frac{4}{15}$ ✔

3 marks

Order of Operations
- Multiplication is done before subtraction

Multiplication of Fractions
- Change mixed fractions to 'top-heavy' fractions $1\frac{1}{5} = \frac{6}{5}$
- Multiply 'top' numbers, multiply 'bottom' numbers $\frac{6}{5} \times \frac{1}{3} = \frac{6 \times 1}{5 \times 3} = \frac{6}{15}$

NOTES: 1·2 page 8

Subtraction of Fractions
- A Common denominator is needed. 3 and 15 both divide exactly into 15.
- $\frac{8}{3} = \frac{8 \times 5}{3 \times 5} = \frac{40}{15}$ multiply 'top' and 'bottom' by 5
- Change back to mixed fractions

NOTES: 1·2 page 9

Q3 $Q = R - S$

$= 3a^2 - b^2$

$= 3 \times (-2)^2 - 3^2$ ✔

$= 3 \times 4 - 9$

$= 12 - 9$

$= 3$ ✔

2 marks

Substitution
- Replace R by $3a^2$ and S by b^2
- Replace a by -2 and b by 3

NOTES: 4·1 page 30

Calculation
- Squares are calculated first:
  $(-2)^2 = -2 \times (-2) = 4$ and $3^2 = 3 \times 3 = 9$
- Multiplication is done before subtraction

NOTES: 1·3 page 9

Q4 $\dfrac{7}{x} - \dfrac{3}{x+1}$

$$= \frac{7(x+1)}{x(x+1)} - \frac{3x}{x(x+1)} \quad \checkmark$$

$$= \frac{7(x+1) - 3x}{x(x+1)}$$

$$= \frac{7x + 7 - 3x}{x(x+1)}$$

$$= \frac{4x + 7}{x(x+1)} \quad \checkmark$$

2 marks

Subtraction of Algebraic Fractions
- A common denominator is needed: multiply x and $x+1$ together
- Always multiply the 'top' and the 'bottom' by the same expression:
$$\frac{7}{x} \times (x+1) \times (x+1) \quad \text{and} \quad \frac{3}{x+1} \times x \times x$$

Simplify
- The denominators are subtracted and then simplified

NOTES: 4·5 pages 38 and 39

Q5 Here is the data in order:

2, 2, 3, 6, 8, 11, 12, 13, 19, 20, 23, 26, 27 $\checkmark$

Now divide data into equal groups:

(2, 2, 3, 6, 8, 11) 12 (13, 19, 20, 23, 26, 27)

$\dfrac{3+6}{2} = 4\cdot5$ (Q$_1$)   Median (Q$_2$) $\dfrac{20+23}{2} = 21\cdot5$ (Q$_3$) $\checkmark$ $\checkmark$

Here is a boxplot of the data:

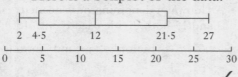

2   4·5   12   21·5   27

0   5   10   15   20   25   30 $\checkmark$

4 marks

The Median
- The data as given is unordered. To calculate the median and quantiles you must order the data from least to greatest.
- The middle value or mean of the middle two values of the ordered data is the median

Quartiles
- If the median is one of the data values then it is not used to calculate the lower and upper quartiles.
- The lower quartile (Q$_1$) is the median of the smaller half of the data.
- The upper quartile (Q$_3$) is the median of the larger half of the data.

Boxplot
- The boxplot uses five statistics: least, lower quartile (Q$_1$), median (Q$_2$), upper quartile (Q$_3$), greatest.
- A scale should be used and the five numbers written on the boxplot as shown.

NOTES: 6·1 pages 56 and 57

Q6 Normal price (100%) has been reduced by 20% to 80%:

80% ⟷ £680 $\checkmark$

1% ⟷ $\dfrac{680}{80}$

100% ⟷ $\dfrac{680}{80} \times 100$ $\checkmark$

$= £850$ $\checkmark$

3 marks

Percentages
- The final price is given *after* a reduction so a 'proportion' method is required.
- Knowing 80% divide by 80 to find 1% and multiply by 100 to find 100%

NOTES: 1·5 page 12

Q7 For the dodecahedral dice of the 12 possible numbers 4 are multiples of 3: 3, 6, 9, 12

Probability $= \dfrac{4}{12} = \dfrac{1}{3}$ ✓

For the Octahedral dice of the 8 possible numbers only 2 are multiples of 3: 3, 6

Probability $= \dfrac{2}{8} = \dfrac{1}{4}$ ✓

Since $\dfrac{1}{3} > \dfrac{1}{4}$ the dodecahedral dice gives a better chance ✓

3 marks

Understanding Probability
- Probability of an event =

  Number of outcomes that make the event happen (favourable outcomes)
  ──────────────────────────
  Total number of outcomes

NOTES: 6·3 page 62

Calculating Probability
- Four 'favourable' outcomes out of a total of 12 possible outcomes (Dodecahedral)
- Two 'favourable' outcomes out of a total of 8 possible outcomes (Octahedral)

Comparison
- It is essential you state the comparison i.e. $\dfrac{1}{3} > \dfrac{1}{4}$ along with your conclusion

Q8(a) This gives:

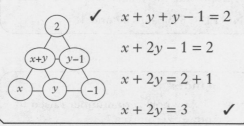

✓ $x + y + y - 1 = 2$

$x + 2y - 1 = 2$

$x + 2y = 2 + 1$

$x + 2y = 3$ ✓

2 marks

Understanding the Pattern
- The example A = B + C tells you that the top circle is the sum of the two below it: $\textcircled{x} + \textcircled{y}$ and $\textcircled{y} + \textcircled{-1}$

Set up an Equation
- Following the pattern for the top triangle leads to an equation:
  $\textcircled{x + y} + \textcircled{y - 1} = \textcircled{2}$

Simplify the Equation
- Letters on the left, numbers on the right

Q8(b) This gives:

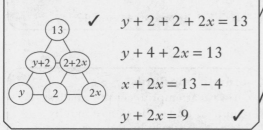

✓ $y + 2 + 2 + 2x = 13$

$y + 4 + 2x = 13$

$x + 2x = 13 - 4$

$y + 2x = 9$ ✓

2 marks

Simultaneous Equations
- You will recognise this from having two equations and two unknowns (x and y)
- Line up the equations x's under x's and y's under y's with numbers on the right

Q8(c) Solve simultaneous equations ✓

$\left.\begin{array}{l} x + 2y = 3 \\ 2x + y = 9 \end{array}\right\} \begin{array}{l} \times 2 \longrightarrow 2x + 4y = 6 \\ \longrightarrow 2x + y = 9 \end{array}$

 Subtract $ 3y = -3$

 so $y = -1$ ✓

Put $y = -1$ into $2x + y = 9$

so $2x - 1 = 9 \Rightarrow 2x = 10$

$\Rightarrow x = 5$ ✓

3 marks

Calculation of values
- Aim for one equation with one unknown
- Substitute the first calculated value back into either of the two equations
- Check using the other equation
  Use $y = -1$ and $x = 5$ in $x + 2y$ to get
  $5 + 2 \times (-1) = 3$

NOTES: 4·4 page 36

Q9 The amplitude is 5 so
$p = 5$ ✓
From 0° to 360° there
would be 3 cycles so $q = 3$ ✓

2 marks

Amplitude
- $y = \sin x°$ has a max or min value of 1 or −1 so $y = 5 \sin x°$ has a max or min value of 5 or −5, true for this graph

Period
- $y = \sin x°$ has period 360° − 1 cycle every 360°. So $y = \sin 3x°$ has period 120° − 3 cycles every 360°; true for this graph.

NOTES: 3·4 pages 26 and 27

Q10 A linear equation:
$$y = px + q$$

gradient y-intercept
(positive   (negative
$p > 0$) $q < 0$)

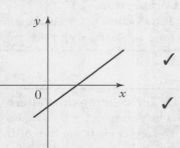

✓
✓

2 marks

Gradient
- $y = px + q$ is "$y = mx + c$" with different letters
- a positive gradient gives an uphill graph

y-intercept
- A negative y-intercept means the graph cuts the y-axis below the x-axis

NOTES: 5·2 pages 47 and 48

Q11(*a*) $8^{\frac{2}{3}} - 8^0$

$= (\sqrt[3]{8})^2 - 1$ ✓

$= 2^2 - 1 = 4 - 1 = 3$ ✓

2 marks

Zero index
- $a^0 = 1$ Any non-zero number raised to the power zero gives 1.

Fractional indices
- $a^{\frac{m}{n}}$ ←power ←root so $a^{\frac{m}{n}} = (\sqrt[n]{a})^m$ and in this case: $8^{\frac{2}{3}}$ ←squared (power 2) ←cube root

NOTES: 4·6 pages 41 and 42

Q11(*b*) $\sqrt{2} + \sqrt{8}$

$= \sqrt{2} + \sqrt{4 \times 2}$

$= \sqrt{2} + \sqrt{4} \times \sqrt{2}$ ✓

$= \sqrt{2} + 2\sqrt{2} = 3\sqrt{2}$ ✓

2 marks

Simplifying Surds
- Always attempt to get the smallest number under the square root by factorising out squares: 4, 9, 16,…

Adding Surds
- Compare $\sqrt{2} + 2\sqrt{2} = 3\sqrt{2}$ with
$$x + 2x = 3x$$

NOTES: 4·6 page 40 and 41

Q12 In 1 hour the tip of the minute hand travels the whole circumference of the clock face ✓

so $\pi D = 27$ ✓

$\Rightarrow D = \dfrac{27}{\pi}$ ✓

The diameter, D, of the clock face is the same length as the side of the square frame

so Perimeter $= 4 \times D$

$= 4 \times \dfrac{27}{\pi}$ ✓

$= \dfrac{108}{\pi}$ metres

4 marks

Strategy
- Evidence that you knew to use the Circumference formula to set up an equation will gain you the strategy mark.

Set up equation
- $C = 2\pi r$ or $C = \pi D$: the second version is the best one for this question

NOTES: 2·4 page 17

Change the subject
- Divide both sides by π

Calculation
- 4 sides make up the perimeter, so $\times 4$
- The working $4 \times \dfrac{27}{\pi}$ is essential especially since the answer $\dfrac{108}{\pi}$ is given.

Practice Exam A: Paper 2 Worked Answers

Q1 Use $T = \dfrac{D}{S}$ ✓

so $T = \dfrac{5\cdot6 \times 10^7}{120}$ ✓

$= 466666\cdot66\ldots$ hours

$= \dfrac{466666\cdot66\ldots}{24}$ days ✓

$= 19444\cdot44\ldots$

$= 1\cdot9 \times 10^4$ days ✓
(to 2 sig figs)

4 marks

D,S,T Formula
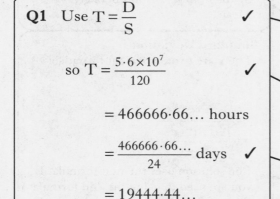
You are asked to find the Time taken. Covering T in the triangle gives $\dfrac{D}{S}$, the Distance divided by the Speed.

Calculator
- $5\cdot6 \times 10^7$ is entered into your calculator like this:

Conversion
- Change hours to days: divide by 24.

Scientific notation
- Convert your answer to scientific notation:

$$a \times 10^n$$

a number between 1 and 10 in this case 1·9   The decimal point in 1·9 moves 4 places to the right so $n = 4$

NOTES: 1·4 page 10

Q2(*a*) Compare $y = mx + c$ ✓
m = gradient

$$= \frac{245 - 120}{5} = \frac{125}{5}$$

$$= 25 \quad ✓$$

"y-intercept" is 120 ✓

Equation is $C = 25t + 120$

3 marks

Linear equation
- You will gain a mark for evidence if you realise "$y = mx + c$" is involved

Gradient
- Gradient $= \dfrac{\text{distance up or down}}{\text{distance along}}$
- Read the scales carefully, they are not the same on the two axes.

Correct Equation
- C is used instead of y and t is used instead of x. $y = 25x + 120$ will not gain the final mark

NOTES: 5·2 page 48

Q2(*b*) $C = 282·5$ ✓

so $25t + 120 = 282·5$

$\Rightarrow 25t = 282·5 - 120$
$\qquad = 162·5$

$\Rightarrow t = \dfrac{162·5}{25} = 6·5$ ✓

The conference lasted $6\frac{1}{2}$ hours. ✓

3 marks

Strategy
- Knowing to use C = 282·5 in the equation will gain you the strategy mark

Solving the equation
- Setting up the correct equation and solving it correctly gain you two more marks.

Q3 $\overline{x}$ (mean) $= \dfrac{\Sigma x}{n}$

$$= \frac{7 + 10 + 8 + 9 + 13 + 7 + 10 + 8}{8}$$

so $\overline{x} = \dfrac{72}{8} = 9$ ✓

x	$x - \overline{x}$	$(x - \overline{x})^2$
7	−2	4
10	1	1
8	−1	1
9	0	0
13	4	16
7	−2	4
10	1	1
8	−1	1

$s = \sqrt{\dfrac{\Sigma (x - \overline{x})^2}{n - 1}}$

$= \sqrt{\dfrac{28}{7}}$

$= \sqrt{4}$ ✓

$= 2$

$\Sigma (x - \overline{x})^2 = 28$ ✓

Mean = 9
standard deviation = 2 ✓

4 marks

The Mean
- The total of the numbers divided by the number of numbers gives the Mean.

Standard Deviation
- There are two equivalent formulae:

$$s = \sqrt{\frac{\Sigma (x - \overline{x})^2}{n - 1}} \text{ and}$$

$$s = \sqrt{\frac{\Sigma x^2 - (\Sigma x)^2 / n}{n - 1}}$$

The solution uses the first formula. If you are used to using the 2nd formula then stick with what you are used to!

- Use a table of values when calculating as it is easier to read and leads to fewer errors

- Working is expected and you will lose marks for not showing working. If you know how to calculate S on your calculater you should do this to check your calculation is correct.

NOTES: 6·2 pages 60 and 61

Q4 Reduction factor is $0\cdot 94$ ✓
and is applied three times.

So Final altitude
$= 340 \times 0\cdot 94 \times 0\cdot 94 \times 0\cdot 94$ ✓

$= 282\cdot 39\ldots \doteqdot 282$ km ✓
(to nearest km)

3 marks

Reduction Factor
• Reducing by 6% is equivalent to calculating 94% i.e. multiplying by $0\cdot 94$

NOTE: $1\cdot 5$ page 11

Time Calculation
• Shorter version is $340 \times 0\cdot 94^3$

Rounding
• Unless accuracy is stated in the question any correct rounding is acceptable.

Q5 Use the sine rule ✓✓

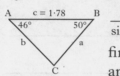

$\dfrac{a}{\sin A}$✓$= \dfrac{b}{\sin B}$✓$= \dfrac{c}{\sin C}$

first calculate angle C

Angle $C = 180° - (46° + 50°)$
$= 180° - 96° = 84°$ ✓

so $\dfrac{b}{\sin 50°} = \dfrac{1\cdot 78}{\sin 84°}$ ✓

$\Rightarrow b = \dfrac{1\cdot 78 \times \sin 50°}{\sin 84°}$

$= 1\cdot 371\ldots \doteqdot \underline{1\cdot 37 \text{ metres}}$ ✓
(to 3 sig figs)

4 marks

Strategy
• If you know two angles and one side then you should use the Sine rule

Angle Sum
• The angle sum in a triangle is 180°.

Substitution
• Correctly substituting the values into the Sine Rule will gain you a mark

Calculation
• Always make sure your calculator is in **degree** mode. Make sure you know how to do this on your calculator.

NOTES: $3\cdot 3$ pages 22 and 23

Q6 The large triangle is isosceles

So angle A = angle C ✓

angle $C = \dfrac{1}{2}(180° - 45°)$

$= \dfrac{1}{2} \times 135° = 67\cdot 5°$ ✓

Now use "SOHCAHTOA" in triangle ABC ✓

$\sin 67\cdot 5° = \dfrac{80}{AC}$ ✓

$\Rightarrow AC \times \sin 67\cdot 5° = 80$

$\Rightarrow AC = \dfrac{80}{\sin 67\cdot 5°} = 86\cdot 59\ldots$

$\doteqdot \underline{\underline{86\cdot 6 \text{ cm}}}$ ✓
(to 3 sig figs)

5 marks

Strategy
• Evidence that you knew to use "SOHCAHTOA" in triangle ABC by finding angle C would gain you the strategy mark

Angle Sum
• Angle A and angle C between them make up 135° of the 180° angle sum in the large triangle

Correct Ratio
• Working from angle C: the opposite AB is known and the hypotenuse AC is required, so use "SOH" i.e. sin

Correct equation
• $\sin 67\cdot 5 = \frac{80}{AC}$ or its equivalent will gain you a further mark

Calculation
• Multiply both sides of the equation by AC and then divide both sides by $\sin 67\cdot 5°$

NOTES: $3\cdot 1$ page 20

Q7   Use the cosine rule ✓
$$b^2 = a^2 + c^2 - 2ac \cos B$$

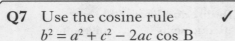

$$= 18^2 + 10\cdot5^2 - 2 \times 18 \times 10\cdot5 \times \cos 95° ✓$$
$$= 467\cdot19\ldots$$
so $b = \sqrt{467\cdot19\ldots} = 21\cdot61\ldots$ ✓

so AC $\doteqdot 21\cdot6$ cm ✓
(to 1 dec. place)

4 marks

Strategy
- Two sides and the included angle are given so the Cosine Rule is used

Substitution
- Correct substitution of the values into the formula will gain you a mark.

Calculation
- In most calculators just key in the complete calculation as written in the solution

Rounding
- 1 mark is allocated for correct rounding

NOTES: 3·3 page 23

Q8   Triangle BCD (see diagram) is isosceles with BC = BD = x ✓

By Pythagoras' Theorem:

$$x^2 + x^2 = 5^2 ✓$$
$$\Rightarrow 2x^2 = 25$$
$$\Rightarrow x^2 = \frac{25}{2} = 12\cdot5 ✓$$

So $x = \sqrt{12\cdot5}$. AC is a radius so AC = 5 ✓
$$AB = AC - BC = 5 - \sqrt{12\cdot5}$$
$$= 5 - 3\cdot53\ldots = 1\cdot464\ldots$$

So AB $\doteqdot 1\cdot46$ metres
(to 3 sig figs) ✓

5 marks

Strategy
- Constructing the right-angled triangle and using Pythagoras' Theorem is the crucial strategy

Pythagoras' Theorem
- Naming BC and BD with letter x helps.

Strategy
- Realisation that AC is a radius, length 5 m, is the 2nd crucial strategy step

Calculation
- Never round answers until the final answer

Rounding
- Rounding to 3 significant figures is essential

NOTES: 2·4 page 19

Q9

Volume = Area of end × length ✓

Use Area = $\frac{1}{2} ab \sin C$

$$= \frac{1}{2} \times 1\cdot4 \times 1\cdot6 \times \sin 50° ✓$$
$$= 0\cdot857\ldots ✓$$

Volume = $0\cdot857\ldots \times 3$ ✓
(area of end)   (length)
$$= 2\cdot573\ldots \doteqdot 2\cdot57\,\text{m}^3 ✓$$

5 marks

Formula
- This formula is true for all prisms

Substitution
- The area formula uses two sides and the included angle

Calculation
- At this stage do not round your answer

Substitution
- Always check the units match—in this case they do

Answer
- Any correct rounding will do for the final answer

NOTES: 3·3 page 22 and 2·1 page 14

Q10 $3\cos x° + 4 = 3$

$\Rightarrow 3\cos x° = -1$

$\Rightarrow \cos x° = -\dfrac{1}{3}$ (negative) ✓

So $x°$ is in the 2nd or 3rd quadrant

1st quadrant angle is given by:

$\cos x° = \dfrac{1}{3} \Rightarrow x = \cos^{-1}\left(\dfrac{1}{3}\right)$

So $x = 70.5$ ✓

This gives:

$x = 180 - 70.5 = 109.5$
(2nd quadrant)

$x = 180 + 70.5 = 250.5$
(3rd quadrant)
(correct to 1 dec. place) ✓

3 marks

Rearrangement
- You solve this equation as you would solve $3c + 4 = 3$ to get $c = -\dfrac{1}{3}$

1st quadrant angle
- Initially ignore the negative sign and use $\boxed{\cos^{-1}}$ button with $\left(\dfrac{1}{3}\right)$ to get an angle in the 1st quadrant—between $0°$ and $90°$

The Quadrants
- Use $\dfrac{S\,|\,A}{T\,|\,C}$ or [graph] to determine

  the quadrants: cosine is negative in 2nd and 3rd quadrants

- 2nd quadrant: $180 - $ (1st quad angle)
  3nd quadrant: $180 + $ (1st quad angle)

NOTES: 3·5 page 27

Q11(*a*)

Use Area = length × breadth

Pool: Outside of Path:

Area $= x(x + 2)$ Area $= (x + 2)(x + 6)$
$= x^2 + 2x$ $= x^2 + 8x + 12$
        ✓   ✓

Area of Path =
Area B − Area A

$= x^2 + 8x + 12 - (x^2 + 2x)$

$= x^2 + 8x + 12 - x^2 - 2x$

$= 6x + 12$ ✓

Area of Pool = Area of Path

$\Rightarrow x^2 + 2x = 6x + 12$

$\Rightarrow x^2 + 2x - 6x - 12 = 0$

$\Rightarrow x^2 - 4x - 12 = 0$ ✓

4 marks

Starting
- You will gain a mark for either of the two area expressions appearing

Strategy
- Any correct method for finding an expression for the area of path and then setting up an equation will guarantee the strategy mark

- There are other methods than the one shown in the solution for finding the areas.

Simplification
- Simplifying expressions as you proceed through your solution makes the subsequent working easier.

Equation rearrangement
- Since the answer is given $(x^2 - 4x - 12 = 0)$ it is important that your working is clear.

Q11(*b*) $x^2 - 4x - 12 = 0$

$\Rightarrow (x - 6)(x + 2) = 0$

$\Rightarrow x = 6 \text{ or } x = -2$ ✓

$x = -2$ is not sensible for a length   ✓

So $x = 6$ is only solution.

Pool dimensions: $6 \text{ m} \times 8 \text{ m}$ ✓

3 marks

Solving a Quadratic
- Each factor could be zero: $x - 6 = 0$ or $x + 2 = 0$ giving the two solutions $x = 6$ and $x = -2$

Invalid Solution
- You must state clearly that $x = -2$ makes no sense. There is a mark for this.

Answer
- The dimensions of the pool are x metres by $(x + 2)$ metres with $x = 6$.

NOTES: 4·7 page 42

Practice Exam B: Paper 1 Worked Answers

Q1 $8{\cdot}82 - 2{\cdot}9 \times 3$

$= 8{\cdot}82 - 8{\cdot}7$ ✓

$= 0{\cdot}12$ ✓

2 marks

> **Order of operations**
> • Multiplication is done before subtraction

> **Decimal Subtraction**
> • Line up the decimal points: $\begin{array}{r} 8{\cdot}82 \\ -8{\cdot}7 \\ \hline 0{\cdot}12 \end{array}$

Q2 $2x^2 + 3x - 2$

$= (2x - 1)(x + 2)$ ✓✓

2 marks

> **Factorisation**
> • You should always check your answer by multiplying out using "FOIL"
>
> NOTES: 4·3 page 33

Q3 $\frac{2}{3}\left(1\frac{1}{2} - \frac{1}{3}\right)$

$= \frac{2}{3}\left(\frac{3}{2} - \frac{1}{3}\right)$

$= \frac{2}{3}\left(\frac{9}{6} - \frac{2}{6}\right) = \frac{2}{3} \times \frac{9-2}{6}$ ✓

$= \frac{2}{3} \times \frac{7}{6} = \frac{14}{18} = \frac{7}{9}$ ✓

2 marks

> **Subtraction of Fractions**
> • The lowest common denominator is 6 as this is the smallest number 2 and 3 divide into exactly

> **Multiplication of Fractions**
> • "Multiply across": $\frac{a}{b} \times \frac{c}{d} = \frac{a \times c}{b \times d}$
> • Cancel if possible: in this case divide top and bottom of the fraction by 2 to get $\frac{7}{9}$
>
> NOTES: 1·2 pages 8 and 9

Q4 $x(2x - 1) - x(2 - 3x)$

$= 2x^2 - x - 2x + 3x^2$ ✓✓

$= 5x^2 - 3x$ ✓

3 marks

> **Brackets**
> • Take care with a negative multiplier: $-x(2 - 3x) = -2x + 3x^2$ (2nd term positive)

> **Simplify**
> • Gather 'like terms': $2x^2 + 3x^2$ and $-x - 2x$
>
> NOTES: 4·1 page 30

Q5(*a*) $f(x) = \frac{6}{x}$

so $f(-3) = \frac{6}{-3} = -2$ ✓

1 mark

> **Substitution**
> • From $f(x)$ to $f(-3)$ involves replacing x by the value -3 and calculating the result.
>
> NOTES: 4·1 page 30

Q5(*b*) $f(a) = 3 \Rightarrow \frac{6}{a} = 3$ ✓

so $6 = 3a \Rightarrow a = 2$ ✓

2 marks

> **Equation**
> • $f(a)$ is replaced by $\frac{6}{a}$ for the 1st mark.

> **Solving**
> • Divide both sides by 3.

Q6(a) $y = -\frac{1}{2}x + 3$ ✓ ✓

2 marks

Equation of straight line
- The general equation of a line is:

$$y = mx + c$$

gradient y-intercept

In this case $m = -\frac{1}{2}$ and $c = 3$

NOTES: 5·2 page 48

Q6(b)   *For* A$(-2, a)$
$x = -2$ and $y = a$

Substitute these values in the line equation:

$a = -\frac{1}{2} \times (-2) + 3$ ✓

$\Rightarrow a = 1 + 3 = 4$ ✓

2 marks

Substitution
- Since A$(-2, a)$ lies on the line its x and y coordinates satisfy the equation of the line. So substitution is the strategy here

Calculation
- Negative × Negative = Positive. No calculator is allowed in Paper 1 work.

NOTES: 5·2 page 48

Q7(a)   Let u be the charge on an 'up' quark and let d be the charge on a 'down' quark:

$2u + d = 1$ ✓

1 mark

Setting up an equation
- It is important to be clear what the letters you use are representing. Here it is the value of the electric charge

Setting up an equation
- 'Algebraic' mentioned in a question means that algebra should be used eg. equation solving, letter substitution etc.

Q7(b) $u + 2d = 0$ ✓

1 mark

Strategy
- Two equations with two unknowns (letters) indicates simultaneous equations need to be solved to find the values of the unknowns.

Q7(c)   Solve simultaneous equations   ✓

$\left.\begin{array}{l} 2u + d = 1 \\ u + 2d = 0 \end{array}\right\} \begin{array}{l} \times 2 \longrightarrow \quad 4u + 2d = 2 \\ \longrightarrow \quad u + 2d = 0 \end{array}$ ✓

subtract $\dfrac{3u \qquad = 2}{}$

$\Rightarrow u = \dfrac{2}{3}$ ✓

2 marks

Method
- Line up the equations: in this case 'u's are lined up and 'd's are lined up
- The aim is always to eliminate one of the unknowns, in this case d.

Calculation
- Divide both sides by 3

NOTES: 4·4 page 36

Q8(*a*)   Total no. of pupils in
class 3M1 = 12 + 16 = 28

Probability of choosing a

boy = $\frac{12}{28} = \frac{3}{7}$ ✓

1 mark

Probability
- Careful reading of the table is required to extract the right information
- Probability =

$$\frac{\text{No. of boys in class 3M1}}{\text{total no. of pupils in class 3M1}}$$

Q8(*b*)   Total no. of pupils

= 12 + 16 + 10 + 14 + 14 + 6 = 72

Total no. of boys
= 12 + 10 + 14 = 36

Probability of choosing a

boy = $\frac{36}{72} = \frac{1}{2}$ ✓

1 mark

Probability
- The totals are now different as all the classes have been put into one large group
- Probability = $\dfrac{\text{Total of all boys}}{\text{Total of all pupils}}$

NOTES: 6·3 page 62

Q9   75% were less efficient than
38 mpg   ✓

1 mark

Boxplot
- The following diagram shows the percentage of data in each part of the plot:

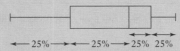

←—25%—→ ←—25%—→ 25% 25%

- 75% of the data occurs below 38 and 25% occurs above.

NOTES: 6·1 page 57

Q10(*a*)   3   :   1
          (nitrogen)   (potassium)
           12 kg   4 kg

His lawn would get 4 kg
potassium.   ✓

1 mark

Ratio
- 3 parts = 12 kg so 1 part = 4 kg
  This is a 'proportion' calculation

Q10(*b*)

   3   :   2   :   1
(nitrogen) (phosphorus) (potassium) ✓
  12 kg   8 kg   4 kg

Total weight
= 12 + 8 + 4 = 24 kg   ✓

3 × 7 kg packets = 21 kg
(3 kg short)

4 × 7 kg packets = 28 kg
(4 kg spare)

So 4 packets are
required   ✓

3 marks

Strategy
- The total weight is required so the weight of Phosphorus has to be calculated

Ratio
- 3 parts = 12 kg, 1 part = 4 kg
  so 2 parts = 2 × 4 kg = 8 kg

Solution
- 24 kg are required. Dividing by 7

gives: $7\overline{)24.^3 0^2 0}$ — $^{3·42...}$. You must however

round up to 4 packets otherwise you will be short. The context of the problem tells you when to round up in this way.

NOTES: 1·5 page 12

Q11(*a*) $S_3 = t_1 + t_2 + t_3$

$\qquad = 1 + 7 + 19 = 27$ ✓

$\qquad S_4 = t_1 + t_2 + t_3 + t_4$

$\qquad = 1 + 7 + 19 + 37$

$\qquad = 64$ ✓

2 marks

Pattern completion
- Completing the S_3 row gains 1 mark and completing the S_4 row gains the 2nd mark
- When solving 'pattern' questions it is important that you compare each line with the previous lines to see how the pattern is building up.

Q11(*b*) $S_3 = 27 = 3^3$

$\qquad S_4 = 64 = 4^3$

$\qquad S_n = n^3$ ✓

1 mark

Generalisation
- Recognition of the 'Cubes' is crucial to solving this question:

$1^3, \quad 2^3, \quad 3^3, \quad 4^3, \quad 5^3, \ldots$

$1, \quad 8, \quad 27, \quad 64, \quad 125, \ldots$

Q11(*c*) $S_{n+1} = (t_1 + t_2 + \ldots + t_n) + t_{n+1}$

$\qquad \Rightarrow S_{n+1} = S_n + t_{n+1}$ ✓

$\qquad \Rightarrow (n+1)^3 = n^3 + t_{n+1}$

$\qquad$ so $t_{n+1} = (n+1)^3 - n^3$ ✓

2 marks

Strategy
- This difficult question requires insight into the relationship between S_n, S_{n+1} and t_{n+1}: adding the first $(n+1)$ terms involves adding the first n terms and then the $(n+1)^{\text{th}}$ term. This, in symbols, is: $S_{n+1} = S_n + t_{n+1}$
- An alternative explanation is that removing the first n terms from the first $(n+1)$ terms will leave only the $(n+1)^{\text{st}}$ term:

$$\underbrace{t_1 + t_2 + t_3 + \ldots + t_n}_{\text{remove these } (S_n)} + t_{n+1}$$

$$\underbrace{\phantom{t_1 + t_2 + t_3 + \ldots + t_n + t_{n+1}}}_{S_{n+1}}$$

Q12(*a*) $9^{-\frac{1}{2}} = \dfrac{1}{9^{\frac{1}{2}}}$ ✓

$\qquad = \dfrac{1}{\sqrt{9}} = \dfrac{1}{3}$ ✓

2 marks

Negative indices
- $a^{-n} = \dfrac{1}{a^n}$ is the general Index Law used

Fractional indices
- $a^{\frac{m}{n}} = \left(\sqrt[n]{a}\right)^m$ is the general Index Law used. This may help:

$a^{\frac{m}{n}} \leftarrow \text{power} \atop \leftarrow \text{root}$ so $a^{\frac{1}{2}} \leftarrow \text{power 1} \atop \leftarrow \text{squre root}$

So $9^{\frac{1}{2}}$ means the square root of 9 to the power 1

Q12(*b*) $\sqrt{t} \times t^2$

$\qquad = t^{\frac{1}{2}} \times t^2$ ✓

$\qquad = t^{\frac{1}{2}+2} = t^{\frac{5}{2}}$ ✓

2 marks

Simplifying indices
- $a^m \times a^n = a^{m+n}$ is the Index law used
- Note $t^{2\frac{1}{2}}$ is not acceptable.

NOTES: 4·6 pages 41 and 42

Q13

Volume of Cuboid

= length × breadth × height

$= 5x \times 2x \times 2x$

$= 20x^3 \text{ cm}^3$ ✓

Blocked Volume

= Area of Triangle × breadth

$= \frac{1}{2} \times AB \times 2x \times 2x$

$= 2 \times AB \times x^2 \text{ cm}^3$ ✓

Blocked Volume $= \frac{1}{4} \times$ Cuboid Volume ✓

$2 \times AB \times x^2 = \frac{1}{4} \times 20x^3$

$\Rightarrow 2 \times AB \times x^2 = 5x^3$

$\Rightarrow AB = \frac{5x^3}{2x^2} = \frac{5x}{2} \text{ cm}$ ✓

4 marks

Volume of a Cuboid
- $5 \times 2 \times 2 = 20$ and $x \times x \times x = x^3$
- You will not lose a mark for forgetting to use the units cm^3

Volume of a Prism
- Volume of Prism
  = Area of end × length
- The triangular end is right-angled so use $\frac{1}{2} \times$ base × height. Base is AB and height = $2x$ cm

Set up Equation
- "One quarter of the volume" is the clue to setting up this equation.

Solving the equation
- Since $x \neq 0$ both sides can be divided by x^2: $\frac{AB \times x^2}{x^2} = \frac{5x^3}{x^2}$

Practice Exam B: Paper 2 Worked Answers

Q1 The yearly increase factor is 1·012. The factor is ✓ applied three times (07–10)

Total amount in 2010 ✓
$= 7.2 \times 1.012 \times 1.012 \times 1.012$
$= 7.46...$ ✓
$\doteq 7.5$ million tonnes ✓
(correct to 2 significant figures).

4 marks

Multiplication Factor
- An increase of 1·2% is the equivalent of calculating 101·2%. Multiplying by 1·012 calculates this amount.

Time calculation
- From 2007 to 2010 involves 3 complete years. Since each year's increase is 1·2% three multiplications by 1·012 are necessary
- A shorter version is 7.2×1.012^3

Calculation
- $\boxed{\wedge}\ \boxed{3}$ These keys mean "raise to the power 3" or "cubing"

Rounding
- When accuracy is mentioned in the question there will be 1 mark allocated for the correct rounding.

NOTES: 1·5 pages 11 and 12

Q2(*a*) Mean

$$= \frac{44+47+38+97+40+52}{6}$$

$$= \pounds 53 \qquad \checkmark$$

1 mark

Mean
- Mean $= \dfrac{\text{total of the numbers}}{\text{the number of numbers}}$

NOTES: 6·2 page 59

Q2(*b*) There are 6 pieces of data so $n = 6$

From part (*a*) $\bar{x} = 53$

x	$x-\bar{x}$	$(x-\bar{x})^2$
44	−9	81
47	−6	36
38	−15	225
97	44	1936
40	−13	169
52	−1	1

$\Sigma(x-\bar{x})^2 = 2448$

$$s = \sqrt{\frac{\Sigma(x-\bar{x})^2}{n-1}}$$

$$= \sqrt{\frac{2448}{5}} \qquad \checkmark$$

$$= \sqrt{489\cdot6}$$

$$= 22\cdot126\ldots$$

$$\doteqdot \pounds 22\cdot13 \qquad \checkmark$$

(to 2 decimal places)

2 marks

Standard Deviation
- You should always double check your calculations and if possible confirm your answer using the STAT mode on your calculator
- It is important that all your working is shown. You will not gain the marks by calculating the standard deviation on your calculator and just writing the answer down

Calculation
- Any correct rounding will gain the mark so long as your calculation is correct.

NOTES: 6·2 pages 60 and 61

Q2(*c*) There was less variation in the amounts spent on a weekday ($s = 8\cdot4$) than on a Saturday ($s = 22\cdot9$, greater than $8\cdot4$) $\qquad \checkmark$

1 mark

Comparison using statistics
- The standard deviation, s, measures the variation of the data about the mean. A greater value of s means more variation, a lesser value means less variation

Q3 Angle CEL $= 320° − 270°$
$$= 50° \qquad \checkmark$$

Angle CLE
$$= 90° − 4° = 86°$$

Angle ECL
$$= 180° − (50° + 86°)$$
$$= 180° − 136°$$
$$= 44° \qquad \checkmark$$

Use the sine rule: $\dfrac{c}{\sin C} = \dfrac{e}{\sin E}$

so $\dfrac{c}{\sin 44°} = \dfrac{11\cdot9}{\sin 50°} \Rightarrow c = \dfrac{11\cdot9 \times \sin 44°}{\sin 50°}$ $\checkmark$

$$= 10\cdot79\ldots$$

So required distance $\doteqdot 10\cdot8$ km $\checkmark$
(correct to 1 dec place)

4 marks

Diagram
- In 'Bearing' questions diagrams are essential
- "due west" gives the clue that the Leven to Elie line LE is at right-angles to the North line
- Bearings are always measured clockwise from the North line

NOTES: 3·2 page 21

Angle sum in triangle
- To use the Sine Rule the 3rd angle must be calculated

Sine Rule
- Multiply both sides by sin 44° to calculate c so that sin 44° will appear at the top of the fraction

Calculation
- Always check that your answer, 10·8 km, appears reasonable in the given context. In this case it is comparable to 11·9 km, the other length, and so seems reasonable.

NOTES: 3·3 pages 22 and 23

Q4(*a*)

Volume of Cylinder $= \pi r^2 \times h$ ✓

In this case

$r = \frac{1}{2} \times 58$

$\quad = 29$ cm, $h = 97$ cm

$V = \pi \times 29^2 \times 97$

$\quad = 256281 \cdot 7 \ldots \ldots$ cm^3 ✓

$\quad = 256 \cdot 2 \ldots$ litres

$\quad \doteqdot 256$ litres ✓
(to the nearest litre)

3 marks

Volume of Cylinder
- This formula is not given and you are expected to know it
- As for any prism the volume is given by the area of the end (πr^2) × the length (h)

Calculation
- Care should be taken that the radius is used in the formula, not the given diameter.

Units
- 1000 cm$^3 = 1$ litre so division by 1000 is necessary to convert to litres

NOTES: 2·1 page 14

Q4(*b*)

$256 \cdot 2 \ldots - 64 = 192 \cdot 2 \ldots$ litres

so $\pi r^2 \times h = 192281 \cdot 7 \ldots$

$\Rightarrow h = \dfrac{192281 \cdot 7 \ldots}{\pi \times 29^2}$

$\quad = 72 \cdot 77 \ldots$ ✓

$\quad \doteqdot 72 \cdot 8$ cm ✓
(to 1 dec place)

2 marks

Strategy
- Using the volume formula with the remaining volume and changing the subject to h will gain you this mark.
- Notice the volume $192281 \cdot 7 \ldots$ is back in cm^3 to allow h to be calculated in cm.

Calculation
- Brackets are essential in the calculator. Calculation: $192281 \cdot 7 \ldots \div (\pi \times 29^2)$
- Using the rounded answer (256 litres) would not be penalised although the calculated height would be $72 \cdot 7$ cm not $72 \cdot 8$ cm

Q5 nth number $= 1 - 3n + 3n^2$

Find n where nth number $= 61$

so $1 - 3n + 3n^2 = 61$ ✓

$\Rightarrow 3n^2 - 3n - 60 = 0$ ✓

$\Rightarrow 3(n^2 - n - 20) = 0$

$\Rightarrow 3(n + 4)(n - 5) = 0$

$\Rightarrow n + 4 = 0$ or $n - 5 = 0$

$\Rightarrow n = -4$ or $n = 5$ ✓

In this case $n = -4$ makes no sense for the position of a number in a sequence

So $n = 5$ is the only solution ✓

61 is the 5th number in the sequence

4 marks

Strategy
- Setting the formula for the nth term of the sequence equal to 61 gains the strategy mark.

Quadratic Equation
- You must recognise this as a quadratic equation and arrange it in the standard form: in this case $an^2 + bn + c = 0$

Solving the equation
- Removal of the common factor first makes the factorising easier
- Both solutions should be given at this stage

NOTES: 4·7 page 42

Solution in context
- Rejection of $n = -4$ should be clear
- You should check that substitution of $n = 5$ in $1 - 3n + 3n^2$ does produce 61

Q6 In triangle ABC

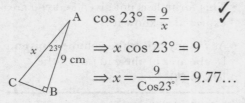

$\cos 23° = \dfrac{9}{x}$ ✓

$\Rightarrow x \cos 23° = 9$

$\Rightarrow x = \dfrac{9}{\cos 23°} = 9.77…$

In triangle ACD

$\sin 35° = \dfrac{y}{9.77..}$ ✓

$\Rightarrow 9.77… \times \sin 35° = y$

$\Rightarrow y = 5.607…$

So CD $\doteq 5.61$ cm ✓
(to 3 sig figs)

4 marks

Strategy
• This question involves a two stage 'SOHCAHTOA': first in triangle ABC and then in triangle ACD.

'SOHCAHTOA'
• Care should be taken when the unknown is on the bottom of the fraction. First multiply by x on both sides of the equation and then divide by cos 23°

• For the 2nd 'SOHCAHTOA' use the full calculator value for AC. Don't round decimals in the middle of a problem only at the end.

Calculation
• Any correct rounding is acceptable since accuracy is not mentioned in the question

NOTES: 3·1 page 20

Q7

Area of triangle $= \dfrac{1}{2} qr \sin P$ ✓

$q = 2·5$ $r = 3·4$

so $4·2 = \dfrac{1}{2} \times 2·5 \times 3·4 \times \sin P$

$\Rightarrow 4·2 = 4·25 \times \sin P$

$\Rightarrow \dfrac{4·2}{4·25} = \sin P$

so $\sin P = 0·9882…$ ✓

$\Rightarrow$ angle $P = \sin^{-1}(0·9882…) = 81·2…°$

so angle $P \doteq 81°$ ✓
(to 2 significant figures)

3 marks

Strategy
• Normally the formula "Area $= \dfrac{1}{2} ab \sin C$" is used to find the Area but in this case the area is given so an equation is set up and solved first for sin P.

Calculation
• The aim is to find a value for sin P

Angle
• Use $\boxed{\sin^{-1}}$ $\boxed{\text{ans}}$ $\boxed{\text{EXE}}$ on your calculator where "ans" uses the previous answer, 0·9882…, on the calculator display

• $\boxed{\text{EXE}}$ and $\boxed{=}$ are the same – it depends on your calculator

NOTES: 3·3 page 22

Q8(*a*) Set $y = 0$ for x-axis intercepts

so $k(x - p)(x - q) = 0$

$\Rightarrow x - p = 0$ or $x - q = 0$

$\Rightarrow x = p$ or $x = q$

From the given graph this gives: ✓

$p = -2$ and $q = 6$ ✓

2 marks

Intercepts
• There is a correspondence between factors and x-axis intercepts:

Factor		Intercept
$x - p$	⟷	$(p, 0)$
$x - q$	⟷	$(q, 0)$

• Since $p = -2$ the factor $x - p = x - (-2) = x + 2$ so $x + 2$ as a factor gives $(-2, 0)$ as an x-axis intercept

NOTES: 4·6 page 42

Q8(*b*)   The equation is
$y = k(x + 2)(x - 6)$ and
from the graph (0, 6) lies
on the parabola

So substitute $x = 0$ and
$y = 6$ into the equation of
the parabola: ✓

$6 = k \times (0 + 2) \times (0 - 6)$

$\Rightarrow 6 = k \times 2 \times (-6)$

$\Rightarrow 6 = k \times (-12)$

$\Rightarrow k = \dfrac{6}{-12} = -\dfrac{1}{2}$ ✓

2 marks

Substitution
• If a point (a, b) lies on a graph then
values $x = a$ and $y = b$ will satisfy the
equation of that graph. Substituting
these values into the equation gives an
equation with only one unknown k in
this case.

Calculation
• Some graphs with
equation
$y = k(x + 2)(x - 6)$ are
shown. They all have
x-intercepts $(-2, 0)$ and $(6, 0)$ and are
all parabolas. Only one graph passes
through $(0, 6)$: $y = -\frac{1}{2}(x + 2)(x - 6)$

Q8(*c*)   The x value that gives
the maximum is half way
between -2 and 6

i.e. $x = \dfrac{-2 + 6}{2} = \dfrac{4}{2} = 2$ ✓

The equation is

$y = -\dfrac{1}{2}(x + 2)(x - 6)$

now substitute $x = 2$

so $y = -\dfrac{1}{2}(2 + 2)(2 - 6)$

$= -\dfrac{1}{2} \times 4 \times (-4) = 8$

The maximum turning
point is (2, 8) ✓

2 marks

x-value
• The graph is symmetrical with axis of
symmetry $x = 2$

• Required value is the mean of -2
and 6

y-value
• The question asks for the coordinates
of a point i.e. (2, 8) not the separate
values.

NOTES: 5·4 page 52

Scale Factor
• If the length scale factor is k
The area scale factor is k^2
The volume scale factor is k^3

Calculation
• For a reduction, the scale factor
should lie between 0 and 1. You
should check that your final answer is
smaller than the given volume of 43
litres.

Rounding
• 1 mark is allocated for correct
rounding to the nearest litre since the
required accuracy is mentioned in the
question

NOTES: 2·2 page 15

Q9   The smaller volume is
required so a reduction
scale factor is needed.

Length scale factor:
$\dfrac{6}{8} = \dfrac{3}{4} = 0.75$

Volume scale factor: 0.75^3 ✓

Required volume
$= 43 \times 0.75^3$
$= 18.14....$ ✓
$\doteq 18$ litres ✓
(to the nearest litre)

3 marks

Q10   AC is a radius
so AC = 2·5 cm

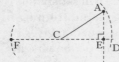

CD is also a
radius
so CD = 2·5 cm ✓

$ED = DF - EF = 5 - 4\cdot8 = 0\cdot2$ cm

and $CE = CD - ED$

$= 2\cdot5 - 0\cdot2 = 2\cdot3$ cm ✓

Use Pythagoras' Theorem in
triangle ACE

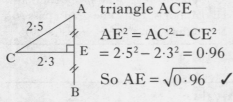

$AE^2 = AC^2 - CE^2$
$= 2\cdot5^2 - 2\cdot3^2 = 0\cdot96$

So $AE = \sqrt{0\cdot96}$ ✓

So the gap $AB = 2 \times AE$

$= 2 \times \sqrt{0\cdot96}$

$= 1\cdot959...$ ✓

$\doteqdot 2\cdot0$ cm (to 1 dec place)

4 marks

Strategy
• The construction of triangle ACE and the use of Pythagoras' Theorem are the essential steps in this solution

Use of Radius
• All radii in a circle are equal. The diameter in this case is 5 cm so radius $= \frac{1}{2} \times 5$ cm

Calculation
• The symmetry of the diagram (CE is an axis of symmetry) means that the gap AB is double the length AE

Answer
• The value 2·0 cm seems reasonable in this context. The value of $\sqrt{0\cdot96}$ should not be rounded before the final answer is reached

• Any correct rounding is acceptable in the final answer as accuracy is not mentioned in the question

NOTES: 2·4 page 19

Q11(*a*)   Using $S = \dfrac{D}{T}$

Average Speed

$= \dfrac{180}{x}$ m.p.h. ✓

1 mark

D.S.T. formula
• Covering up S reveals $\frac{D}{T}$ as the formula for calculating Average Speed.

Using algebra
• You should be aware that "2 less than x" translates to the algebraic expression $x - 2$

• Note the distance has changed to 60 miles

Q11(*b*)   Again using $S = \dfrac{D}{T}$

"2 hours less" gives $x - 2$ hours

So Average speed $= \dfrac{60}{x-2}$ m.p.h. ✓

1 mark

Q11(*c*)

The two average speeds are equal so:

$$\frac{180}{x} = \frac{60}{x-2} \qquad \checkmark$$

$$\Rightarrow 180(x - 2) = 60x$$

$$\Rightarrow 180x - 360 = 60x$$

$$\Rightarrow 120x = 360$$

$$\Rightarrow x = 3 \qquad \checkmark$$

1st stage took $x = 3$ hours

2nd stage took $x - 2 = 1$ hour

Total journey took 4 hours $\qquad \checkmark$

3 marks

Set up equation
- The strategy here is to equate the two expressions for the average speeds.
- You would still gain this mark even if you equated "wrong" expressions

Solving
- Multiply both sides of the equation by x and then by $x - 2$ (or "cross-multiply")

Interpretation
- What does $x = 3$ mean? You must look back to x hours and $x - 2$ hours and substitute.

NOTES: 4·4 page 34

Practice Exam C: Paper 1 Worked Answers

Q1 $1 \cdot 3 \times (4 \cdot 92 + 0 \cdot 08)$

$= 1 \cdot 3 \times 5$ ✓

$= 6 \cdot 5$ ✓

2 marks

Order of Operations
- Calculations in brackets are always done first

Calculation
- No calculator is allowed in Paper 1
- When adding decimals remember to line up the decimal points: $\begin{array}{r} 4 \cdot 92 \\ + 0 \cdot 08 \\ \hline 5 \cdot 00 \end{array}$

Q2 $\frac{2}{3} \div 1\frac{1}{3}$

$= \frac{2}{3} \div \frac{4}{3} = \frac{\frac{2}{3}(\times 3)}{\frac{4}{3}(\times 3)}$ ✓

$= \frac{2}{4} = \frac{1}{2}$ ✓

2 marks

Division of Fractions
- Change mixed fractions to 'top-heavy'
- An alternative method to the one given is to 'invert then multiply' so $\frac{2}{3} \div \frac{4}{3} = \frac{2}{3} \times \frac{3}{4} = \frac{6}{12} = \frac{1}{2}$

Calculation
- Remember to 'cancel down' if possible
NOTES: 1·2 page 8

Q3 $f(x) = x(2 - x)$

$\Rightarrow f(-1) = -1 \times (2 - (-1))$ ✓

$= -1 \times (2 + 1)$

$= -1 \times 3 = -3$ ✓

2 marks

Substitution
- for $f(-1)$, every occurence of x is replaced by the value -1. No further calculation is needed to gain the 1st mark.

Calculation
- The 2nd mark here is for the subsequent calculation after the substitution
- Remember: brackets first and that subtracting a negative is the same as adding the positive value

NOTES: 4·1 page 30

Q4 $3 + 2x < 4(x + 1)$

$\Rightarrow 3 + 2x < 4x + 4$ ✓

(now subtract 4 from each side)

$\Rightarrow -1 + 2x < 4x$

(now subtract $2x$ from each side)

$\Rightarrow -1 < 2x$ ✓

$\Rightarrow \frac{-1}{2} < x$

so $x > -\frac{1}{2}$ ✓

3 marks

Brackets
- Both terms are multiplied by 4. A common mistake here would be $4x + 1$.

Simplification
- The 'balancing process' aims to get letters on one side of the inequality sign and numbers on the other side
- Best to avoid $-2x < 1$. The next step here would be divide by -2 and 'swap the sign round': $x > \frac{1}{-2}$

Answer
- Notice if $a < b$ then $b > a$
NOTES: 4·4 page 35

Q5(*a*) $9y^2 - 4$

$\quad\quad = (3y)^2 - 2^2$

$\quad\quad = (3y - 2)(3y + 2)$ ✔

1 mark

Difference of Squares
- The pattern is: $A^2 - B^2 = (A - B)(A + B)$
- You should always check that your answer multiplies out giving the original expression (use 'FOIL')

NOTES: 4·3 page 32

Q5(*b*) $\dfrac{9y^2 - 4}{15y - 10}$

$\quad\quad = \dfrac{(3y - 2)(3y + 2)}{5(3y - 2)}$ ✔

$\quad\quad = \dfrac{3y + 2}{5}$ ✔

2 marks

Factorising
- When simplifying algebraic fractions the 1st step is to factorise both expressions
- 'Hence' appearing in a question means you should use the answer that you got in the previous part of the question: in this case $9y^2 - 4 = (3y - 2)(3y + 2)$

Cancelling
- Any factor eg $3y - 2$ that appears on the top and the bottom can be cancelled.

NOTES: 4·5 page 37

Q6 $P = \dfrac{W}{4A}$

$\quad\quad (\times A)\ (\times A)$

$\quad\quad \Rightarrow PA = \dfrac{W}{4}$ ✔

$\quad\quad (\div P)\ (\div P)$

$\quad\quad \Rightarrow A = \dfrac{W}{4P}$ ✔

2 marks

1st rearrangement
- The subject of the formula, A, in this case appears on the bottom of the fraction i.e. W is divided by A. The inverse process is multiplication so multiply both side of the formula by A.

2nd rearrangement
- A is multiplied by P. The inverse process is division so divide both sides by P.

NOTES: 4·5 pages 39 and 40

Q7 Use the Cosine rule

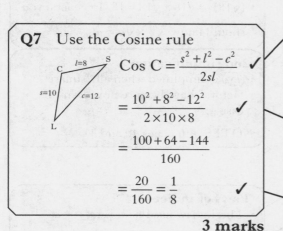

$\quad\quad \text{Cos C} = \dfrac{s^2 + l^2 - c^2}{2sl}$ ✔

$\quad\quad = \dfrac{10^2 + 8^2 - 12^2}{2 \times 10 \times 8}$ ✔

$\quad\quad = \dfrac{100 + 64 - 144}{160}$

$\quad\quad = \dfrac{20}{160} = \dfrac{1}{8}$ ✔

3 marks

Cosine Formula
- The formula as given in the exam is:

$$\cos A = \dfrac{b^2 + c^2 - a^2}{2bc}$$

and you have to be able to adapt this to the particular letters used in the example.

Substitution
- Be very careful with order of the letters. Swapping values eg l and c will give you the wrong answer.

Calculation
- It is important to show all your working since the answer, $\frac{1}{8}$, is given.

NOTES: 3·3 page 24.

Q8

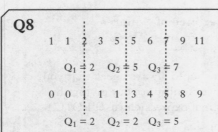

1 1 2 3 5 ⋮ 5 6 7 ⋮ 9 11

$Q_1 = 2$ $Q_2 = 5$ $Q_3 = 7$

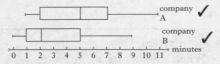

0 0 1 1 1 ⋮ 3 4 5 ⋮ 8 9

$Q_1 = 2$ $Q_2 = 2$ $Q_3 = 5$ ✓

Boxplots:

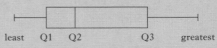

company A ✓

company B ✓

0 1 2 3 4 5 6 7 8 9 10 11 minutes

3 marks

Choice of diagram
- Stem-and-leaf (back to back) would not be appropriate in this case since most of the data is single digit values.
- To calculate Q_1 (Lower quartile), Q_2 (median) and Q_3 (upper quartile) each of the data sets has to be ordered —least to greatest

NOTES: 6·1 page 56

Boxplots
- For comparison purposes the two boxplots are shown on the same scale
- Here is a reminder of how each statistic is used to construct the boxplot:

least   Q1   Q2   Q3   greatest

NOTES: 6·1 page 57

Q9 $f(x) = g(x)$

$\Rightarrow 4 + 3x - x^2 = 10 - 2x$ ✓

$\Rightarrow 0 = 10 - 2x - 4 - 3x + x^2$

$\Rightarrow x^2 - 5x + 6 = 0$ ✓

$\Rightarrow (x - 2)(x - 3) = 0$

$\Rightarrow x - 2 = 0$ or $x - 3 = 0$

$\Rightarrow x = 2$ or $x = 3$ ✓

3 marks

Set up equation
- $f(x)$ is replaced by $4 + 3x - x^2$ and $g(x)$ is replaced by $10 - 2x$

Standard form
- When you recognise a quadratic equation then rearrange it to the form: $ax^2 + bx + c = 0$

Solving Quadratic Equation
- Solving a quadratic equation will not require "the formula" unless the question asks for the roots "to 1 dec. place" or similar

NOTES: 4·7 page 42

Q10 $\sqrt{(\sqrt{18})^2 + (\sqrt{6})^2}$

$= \sqrt{18 + 6} = \sqrt{24}$ ✓

$= \sqrt{4 \times 6} = \sqrt{4} \times \sqrt{6}$

$= 2 \times \sqrt{6} = 2\sqrt{6}$ ✓

2 marks

1st simplification
- $(\sqrt{18})^2 = \sqrt{18} \times \sqrt{18} = 18$. In general you should know: $\sqrt{a} \times \sqrt{a} = a$

2nd simplification
- $\sqrt{24}$ is simplified when all "square factors" have been removed: in this case 4.

NOTES: 4·6 pages 40 and 41

Q11 $x^3 \times (x^{-1})^{-2}$

$= x^3 \times x^{-1 \times (-2)} = x^3 \times x^2$ ✓

$= x^{3+2} = x^5$ ✓

2 marks

Laws of indices
- The 1st law used here is $(a^m)^n = a^{mn}$
- The 2nd law used is $a^m \times a^n = a^{m+n}$

NOTES: 4·6 pages 41 and 42

Q12 gradient $= \dfrac{24-20}{16} = \dfrac{4}{16} = \dfrac{1}{4}$ ✓

The intercept is (0, 20) ✓

Equation is $L = \dfrac{1}{4}W + 20$ ✓

3 marks

Gradient
- Gradient $= \dfrac{\text{distance up (or down)}}{\text{distance along}}$
- Take case when reading the scales as they are not the same on the two axes.

"$y = mx + c$"
- The equation follows the pattern:

$$y = mx + c$$

this is L   gradient $= \frac{1}{2}$ this is W   (0, 20) is the intercept

NOTES: 5·2 page 48

Q13(a) Let k euros be the cost of 1 kg of Kenyan and r euros be the cost of 1 kg of Rwandan

so $2k + 3r = 6\cdot1$ ✓

1 mark

1st equation
- 'algebraic' refers to using letters
- It is important to state clearly what each letter stands for: in this case the cost of 1 kg

Q13(b) $3k + 2r = 5\cdot9$ ✓

1 mark

2nd equation
- Notice that there are no units appearing in the equations
- When simultaneous equations are anticipated it is useful to write the equations in the form: $ax + by = c$ where a, b and c are the given values eg 2, 3, 6·1 or 3, 2, 5·9 and x and y are the variables, in this case k and r.

Q13(c) Solve simultaneous equations:

$2k + 3r = 6\cdot1$ |×3 → $6k + 9r = 18\cdot3$
$3k + 2r = 5\cdot9$ |×2 → $6k + 4r = 11\cdot8$ ✓

Substract: $5r = 6\cdot5$
$\Rightarrow r = 1\cdot3$ ✓

Substitute $r = 1\cdot3$ in $2k + 3r = 6\cdot1$

$\Rightarrow 2k + 3 \times 1\cdot3 = 6\cdot1$

$\Rightarrow 2k + 3\cdot9 = 6\cdot1$

$\Rightarrow 2k = 2\cdot2 \Rightarrow k = 1\cdot1$ ✓

So 1 kg of Kenyan costs 1·10 euros and 1kg of Rwandan costs 1·30 euros

So 4 kg of Kenyan and 1 kg of Rwandan costs $4 \times 1\cdot10 + 1 \times 1\cdot30$ $= 5\cdot70$ euros
5 kg of Blend C costs 5·70 euros ✓

4 marks

Method
- The 'simultaneous equation' strategy will be rewarded with a 'strategy mark'

1st variable
- A further mark is awarded for the correct calculation of one of the variables, either k or r.
- In the working shown 'k' has been eliminated. 'r' could have been eliminated by multiplying the top equation by 2 and the bottom equation by 3 and then subtracting

2nd variable
- Either of the two equations can be chosen for substitution of the 1st value to calculate the 2nd value

Answer
- The question did not ask for just the values of k and r!

NOTES: 4·4 page 36

Practice Exam C: Paper 2 Worked Answers

Q1 N° of letters $= \dfrac{1\cdot05\times10^{-1}}{3\times10^{-8}}$ ✓

$= 35\,00000$

$= 3\cdot5\times10^{6}$ ✓

2 marks

Strategy
- This is a division, it is sometimes useful to try to solve a simpler similar problem to see how the solution is found:
- If each letter were 2 mm wide and the line was 6 mm long there would be 3 letters: 6 divided by 2 gives 3.

Calculation
- $1\cdot05\times10^{-1}$ is entered in the calculator as follows: ⟨1⟩ ⟨·⟩ ⟨0⟩ ⟨5⟩ ⟨×10ˣ⟩ ⟨(−)⟩ ⟨1⟩
- On your calculator, you may have ⟨EXP⟩ not ⟨×10ˣ⟩

NOTES: 1·4 page 10

Q2 It is now $106\tfrac{1}{2}\%$ of the original cost so:

$106\cdot5\% \longleftrightarrow £87\cdot82$ ✓

$1\% \longleftrightarrow \dfrac{£87\cdot82}{106\cdot5}$

$100\% \longleftrightarrow \dfrac{£87\cdot82}{106\cdot5}\times100$ ✓

$= £82\cdot46$ ✓

3 marks

Strategy
- The 'original' price has to be found. It will be 100%. This is now a 'proportion' problem: find 1% then find 100%

Calculation
- It is sometimes best not to calculate intermediate answers eg $£87\cdot82 \div 106\cdot5$ but build up the calculation on paper and then do the calculation at the end. The solution does just that.

NOTES: 1·5 page 12

Q3 $3x^2 - x - 5 = 0$

Compare $ax^2 + bx + c = 0$

$\Rightarrow a = 3,\ b = -1,\ c = -5$

$x = \dfrac{-b \pm \sqrt{b^2 - 4ac}}{2a}$

$= \dfrac{1 \pm \sqrt{(-1)^2 - 4\times3\times(-5)}}{2\times3}$ ✓

$= \dfrac{1 \pm \sqrt{1 + 60}}{6} = \dfrac{1 \pm \sqrt{61}}{6}$ ✓

so $x = \dfrac{1 + \sqrt{61}}{6}$ or $x = \dfrac{1 - \sqrt{61}}{6}$

$x = 1\cdot468\ldots$ or $x = -1\cdot135\ldots$ ✓

$x \doteq 1\cdot5$ or $x \doteq -1\cdot1$ ✓

4 marks

Substitution
- Care should be taken over negative values
- It is safer to state the values of a, b and c that you are going to use. Correct substitution into the formula will gain the 1st mark

Simplification
- Before 'going decimal' it is clearer, and avoids mistakes, to separate the two solutions (roots) as is done in the 3rd last line

Calculation
- Giving a clear indication of each decimal before it is rounded will allow you to gain the last mark for rounding even if your answers are wrong before rounding

Rounding
- Always follow precisely any rounding requests

NOTES: 4·7 page 43

Q4

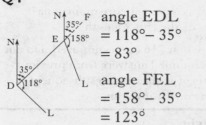

angle EDL ✓
$= 118° − 35°$
$= 83°$ ✓

angle FEL
$= 158° − 35°$
$= 123°$

So angle DEL
$= 180° − 123° = 57°$
angle ELD $= 180° − (83° + 57°)$
$= 180° − 140° = 40°$ ✓

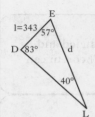

Now use the Sine Rule

$$\frac{d}{\sin D} = \frac{l}{\sin L}$$

$$\Rightarrow \frac{d}{\sin 83°} = \frac{343}{\sin 40°}$$ ✓

$$\Rightarrow d = \frac{343 \times \sin 83°}{\sin 40°} = 529 \cdot 63\ldots$$

$$\doteqdot 530$$ ✓

The distance from Edinburgh to London is approximately 530 km to the nearest km.

5 marks

Strategy
- You need to calculate the three angles in triangle DEL and then use the Sine Rule

1st angle
- Remember bearings are always measured clockwise from the North line

2nd & 3nd angles
- Extending the line DE (EF in the diagram) allows you to move the information at angle D up to angle E (35° is a corresponding angle)

NOTES: 3·2 page 21

The Sine Rule
- Substitution of the values into the Sine Rule will gain you a mark.

Calculation
- Multiply both sides of the equation by sin 83°
- Any correct rounding is acceptable.

NOTES: 3·3 pages 22 and 23

Q5(*a*) Use $A = \pi r^2$
Outside circle diameter
$= 30$ cm $\Rightarrow$ radius $= 15$ cm

Inside circle diameter
$= 25$ cm $\Rightarrow$ radius
$= 12 \cdot 5$ cm

$\begin{matrix} \text{Area of} \\ \text{plastic} \end{matrix} = \begin{matrix} \text{Outside} \\ \text{circle area} \end{matrix} − \begin{matrix} \text{Inside} \\ \text{circle area} \end{matrix}$ ✓

$= \pi \times 15^2 − \pi \times 12 \cdot 5^2$ ✓

$= \pi(15^2 − 12 \cdot 5^2)$

$= \pi(15 − 12 \cdot 5)(15 + 12 \cdot 5)$

$= \pi \times 2 \cdot 5 \times 27 \cdot 5$

$= 215 \cdot 98\ldots$

$\doteqdot 216$ cm^2 ✓

(to the nearest 1 cm^2)

3 marks

Strategy
- Removal of the smaller circle area from the area of the larger outer circle is the crucial idea for the strategy mark.

Substitution
- The given diameter measurements should be halved to give the radius. This is necessary as the area formula $A = \pi r^2$ uses the radius value and not the diameter value.

Calculation
- The use of common factor and difference of squares is not expected— but it is fun!
- The alternative is to enter $\pi \times 15^2 − \pi \times 12 \cdot 5^2$ as written straight into your calculator.

NOTES: 2·4 page 17

Q5(*b*)

$$\text{Volume} = \text{Area of end} \times \text{length}$$
$$= 215 \cdot 98.... \times 100$$
$$= 21598 \cdot 44.... \text{ cm}^3 \quad \checkmark$$

$$\text{Weight} = 21598 \cdot 44... \times 0 \cdot 12$$
$$= 2591 \cdot 81... \text{ g}$$
$$= 2 \cdot 591... \text{ kg}$$
$$\doteqdot 2 \cdot 6 \text{ kg} \quad \checkmark$$
(to 2 sig figs)

2 marks

Volume
- This is a typical "prism formula": area of the end × length
- Note that 216 cm² is not used. Do not use rounded answers from previous parts of a question in subsequent parts of the question.

Weight
- Rounding is not crucial and conversion to kg not essential.
NOTES: 2·1 page 14

Q6(*a*)

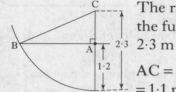

The radius of the fuselage is 2·3 m

$$AC = 2 \cdot 3 - 1 \cdot 2$$
$$= 1 \cdot 1 \text{ m (see diagram)} \quad \checkmark$$

Use Pythagoras' Theorem in triangle ABC:

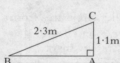

$$AB^2 = BC^2 - AC^2$$
$$= 2 \cdot 3^2 - 1 \cdot 1^2$$
$$= 4 \cdot 08$$

so $AB = \sqrt{4 \cdot 08} = 2 \cdot 019... \checkmark$

By symmetry: width of floor $= 2 \times AB$

so $x = 2 \times 2 \cdot 019... = 4 \cdot 039... \checkmark$
$\Rightarrow x \doteqdot 4 \cdot 04$ metres.

4 marks

Strategy
- Triangle ABC is right-angled and therefore Pythagoras' Theorem can be used.

Pythagoras' Theorem
- Notice that CB is a radius and that CA is part of a radius. Any radius in this circle has a length $\frac{1}{2} \times 4 \cdot 6 = 2 \cdot 3$ m

Calculation
- AB is one of the smaller sides in triangle ABC so the calculation involves a subtraction
- Remember that rounding off decimal answers should only be done at the end of the problem when the final answer is given

Solution
- By symmetry AB is half the width of the compartment floor.

Q6(*b*) By symmetry (see diagram) the ceiling is as far above C as the floor is below C.

$$\text{Height} = 2 \times 1 \cdot 1$$
$$= 2 \cdot 2 \text{ metres} \checkmark$$

1 mark

Solution
- Two equal and parallel chords in a circle will be the same distance from the centre because the diagram is symmetric.

NOTES: 2·4 page 19

Q7   9 parts   :   2 parts
(honey)   (cocoa)

Try to use all the cocoa:

So 2 parts cocoa weighs 12 kg

So 1 part weighs 6 kg

So 9 parts honey weighs $9 \times 6 = 54$ kg

There is not enough honey.   ✓

Try to use all the honey:

So 9 parts honey weighs 49·5 kg

So 1 part weighs $49·5 \div 9 = 5·5$ kg

So 2 parts cocoa weighs $2 \times 5·5 = 11$ kg   ✓

leaving 1 kg cocoa unused.

Total weight used $= 49·5 + 11 = 60·5$ kg

So 60 1–kg jars can be made.   ✓

3 marks

Strategy

- In a question like this you will know that some of the ingredients will be left over when the mixing is complete. This is because, in this case, $49\frac{1}{2} : 12$ is not the same as 9 : 2. The strategy is to determine, with working, which ingredient will be left over.

- Alternative working here might be that 9 : 2 is the same as 54 : 12 (multiplying by 6) and comparing $49\frac{1}{2} : 12$

Calculation

- Calculation of the amount of cocoa required, 11 kg, to go with $49\frac{1}{2}$ kg honey will gain you this work

Answer

- Round down not up—you need complete jars!

NOTES: 1·5 page 12

Q8   For the points of intersection solve: $\cos x° = 0·2$ ✓

$x°$ is in the 1st or 4th quadrants

1st quadrant angle is given by: $x = \cos^{-1}(0.2) \doteq 78·5°$ ✓

4th quadrant angle is given by: $x = 360 - 78·5 = 281·5°$

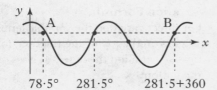

78·5°   281·5°   281·5+360

The x-coordinate of A is 78·5

From the graph above B lies 1 cycle on from the 4th quadrant intersection

The x-coordinate of B is $281·5 + 360 = 641·5$ ✓

3 marks

Strategy

- At each of the points of intersection the y-coordinate is 0·2 and the x-coordinate satisfies $\cos x° = 0·2$. The appearance of the equation $\cos x° = 0·2$ and an attempt to solve it will gain you this mark.

1st quadrant angle

- Using $\boxed{\cos^{-1}}$ on your calculator will always give you the 1st quadrant angle (don't use this key with a negative value)

Other angles

- You should identify the 4 quadrants on the graph that is given:

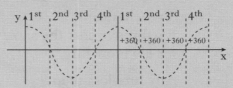

It should then be clear that the x-coordinate of B is given by the 4th quadrant value + 360°

- The 4th quadrant value is obtained using this diagram:

NOTES: 3·5 page 27

Q9(*a*) For 2 hours:

Earthmove: $64 + 2 \times 30 = £124$

(delivery) (2 hours)

Trenchers: $28 + 2 \times 34 = £96$

(delivery) (2 hours) ✓

1 mark

Calculation
- Both values correct will gain you this mark

Q9(*b*) For 20 hours:

Earthmove: $64 + 20 \times 30$
$= £664$

Trenchers: $28 + 20 \times 34$
$= £708$ ✓

So Earthmove is cheaper in this case by £44

1 mark

Comparison
- The two calculations have to be clearly shown
- A comparison should be made between the two costs eg "£44 cheaper" or "£664 is less than £708" etc

1st formula
- either one of the two expressions correctly stated will gain you the 1st mark here.

2nd formula
- Two correct expressions: two marks gained.

Q9(*c*) For *n* hours:

Earthmove: $64 + n \times 30$
$= £30n + 64$ ✓

Trenchers: $28 + n \times 34$
$= £34n + 28$ ✓

2 marks

Strategy
- From parts (*a*) & (*b*) you should have realised that at some point Trenchers becomes the more expensive option having been cheaper when the number of hours was low.
- Using the formulae you worked out in (*c*) and setting them equal to each other would gain you this mark.

NOTES: 4·4 page 34.

Q9(*d*) For a 2 hour hire Trenchers are cheaper but not for a 20 hour hire. To find the 'cross-over value':

Solve $34n + 28 = 30n + 64$
$\Rightarrow 4n = 36 \Rightarrow n = 9$ ✓

Trenchers and Earthmove cost the same for a 9 hour hire.

Any hire less than 9 hours is cheaper with Trenchers. ✓

2 marks

Solution & Interpretation
- What does $n = 9$ mean in the context of the question? A statement is necessary

Variation Formula
- For direct variation the variable letter, L, goes on the top of the fraction, for inverse variation the letter, T, goes on the bottom.
- In general if A varies directly as B then $A = k \times B$. If A varies inversely as B
Then $A = k \times \frac{1}{B}$ where K is a constant

Q10(*a*) $D \propto \dfrac{\sqrt{L}}{T^2}$

so $D = k \times \dfrac{\sqrt{L}}{T^2}$ ✓

where *k* is a constant.

1 mark

Q10(*b*) Replace L by 4L and T by 2T:

So $D = k \times \dfrac{\sqrt{L}}{T^2}$

becomes: ✓

$\dfrac{k \times \sqrt{4L}}{(2T)^2} = \dfrac{k \times 2\sqrt{L}}{4T^2}$

$= \dfrac{2}{4} \times \dfrac{k\sqrt{L}}{T^2}$

$= \dfrac{1}{2}D$

So the Diameter, D, is halved. ✓

2 marks

Strategy
- The method here is to take the formula for D and replace the right-hand side of the formula by a new expression: one where each of the letters is replaced by the results of the changes described in the question. This new expression is then simplified and compared with the original expression for D.

Answer
- The result of the simplification should be described—D is now multiplied by $\frac{1}{2}$.

NOTES: 4·8 pages 44 and 45

Pattern Extension
- The order of writing the line cannot be altered.

Q11(*a*)
  4th line: $8 + 1 = 4 \times 6 - 5 \times 3$ ✓

1 mark

1st expression
- For finding one of $2n + 1$ or $n(n + 2)$ or $(n + 1)(n - 1)$ you will gain the 1st mark here

Q11(*b*)

1st line: $2 + 1 = 1 \times 3 \quad - 2 \times 0$

2nd line: $4 + 1 = 2 \times 4 \quad - 3 \times 1$

3rd line: $6 + 1 = 3 \times 5 \quad - 4 \times 2$

4th line: $8 + 1 = 4 \times 6 \quad - 5 \times 3$

$\vdots$ ✓

*n*th line: $2n + 1 = n(n + 2) - (n + 1)(n - 1)$

  ✓

2 marks

Remaining expressions
- Careful use of brackets is essential: $n \times n + 2 \ (= n^2 + 2)$ is not the same as $n(n + 2)$
  $n + 1 \times n - 1 \ (= 2n - 1)$ is not the same as $(n + 1)(n - 1)$
- Always check: eg substituting $n = 3$ should produce the 3rd line.

Simplification
- Again great care should be taken with brackets especially expanding $(n + 1)(n - 1)$ because $-(n^2 - n + n - 1)$ is not the same as $-n^2 - n + n - 1$
- Proving $2n + 1 = n(n + 2) - (n + 1)(n - 1)$ using 'algebra' means that all lines are true for instance $n = 10$ gives:
  $20 + 1 = 10 \times 12 - 11 \times 9$ which is line 10.

NOTES: 4·2 page 31

Q11(*c*) Simplify the right-hand side of the *n*th line:

$n(n + 2) - (n + 1)(n - 1)$

$= n^2 + 2n - (n^2 - n + n - 1)$

$= n^2 + 2n - n^2 + n - n + 1$

$= 2n + 1$ ✓

This is the same as the left-hand side of the *n*th line.

So the pattern always holds.

1 mark

Q12 triangles ABC and ADE are similar (equiangular) ✓

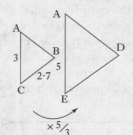

$$AE = AC + CE$$

$$= 3 + 2$$

$$= 5 \text{ m}$$

The enlargement scale factor is $\frac{5}{3}$

so $DE = \frac{5}{3} \times 2 \cdot 7 = 4 \cdot 5$ m ✓

Use Pythagoras' Theorem in triangle DEH: ✓

$$EH^2 = ED^2 - DH^2$$

$$= 4 \cdot 5^2 - 2 \cdot 5^2$$

$$= 14 \quad ✓$$

so $GF = EH = \sqrt{14} = 3 \cdot 714 \ldots$

The gap between the vertical ✓ posts is $3 \cdot 71$ m (to 3 sig figs)

5 marks

Strategy
- Similar triangles calculation to find DE followed by Pythagoras' Theorem in triangle DEH

Similar Triangles
- Enlargements have scale factor greater than 1
- You should always 'disentangle' the two similar triangles, drawing them separately and writing in each of the lengths that you know

NOTES: 2·2 page 15

Pythagoras' Theorem
- Finding a smaller side involves a subtraction

Calculation
- Remember to use $\boxed{x^2}$ key in this calculation

Answer
- Always check that your final answer seems reasonable in the context of the question. In this case $3 \cdot 71$ m compares well to just over 5 m for the 'king post' so seems reasonable.

NOTES: 2·3 page 16

Practice makes perfect!

Your **Practice Papers for SQA Exams** with worked answers give you the most effective exam revision resource available. There are 30 titles to choose from, each packed with full practice exams and answer sections that contain mark allocations and loads of helpful hints and top exam tips from experienced authors and examiners.

Intermediate 2 & Higher Practice Papers
(see inside front cover for Standard Grade & Intermediate 1 titles)

 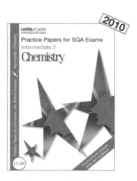   

Intermediate 2 Biology   Intermediate 2 Chemistry   Intermediate 2 English   Intermediate 2 Maths Units 1, 2, 3   Intermediate 2 Maths Units 1, 2, Apps

  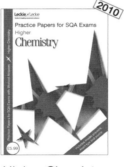  

Higher Biology   Higher Business Management   Higher Chemistry   Higher Computing   Higher English

Higher Geography   Higher History   Higher Human Biology   Higher Maths   Higher Physics

How to buy...

All Practice Papers for SQA Exams are available to buy from your local bookshop. You can also order our books direct via www.leckieandleckie.co.uk or phone us on 0870 460 7662.

Practice Papers for SQA Exams

Standard Grade | Credit

Mathematics

✕ Ken Nisbet ✕

Practice makes perfect!

- Completely new Practice Papers
- Fully worked out answer sections to show students not just the answer but how to reach it
- Packed full of practical exam hints and tips from experienced teachers and examiners
- Question papers look and feel just like the SQA exam to mirror the exam experience
- Includes a helpful topic index so you can practise specific subjects

The most effective revision resource available for Scottish students wanting to practise exam questions and understand how to achieve the best grade.

Scotland's leading educational publishers

Leckie & Leckie
4 Queen Street, Edinburgh EH2 IJE
T: **0131 220 6831** F: **0131 225 9987**
E-mail: **enquiries@leckieandleckie.co.uk**
Web: **www.leckieandleckie.co.uk**

PEFC™
PEFC/16-33-321

ISBN 978-1-84372-773-6

9 781843 727736

Leckie & Leckie is a division of Huveaux plc

Practice Papers for SQA Exams with Worked Answers ✕ Standard Gr

KT-950-370

caa CYMRU

cyfres lobsgows

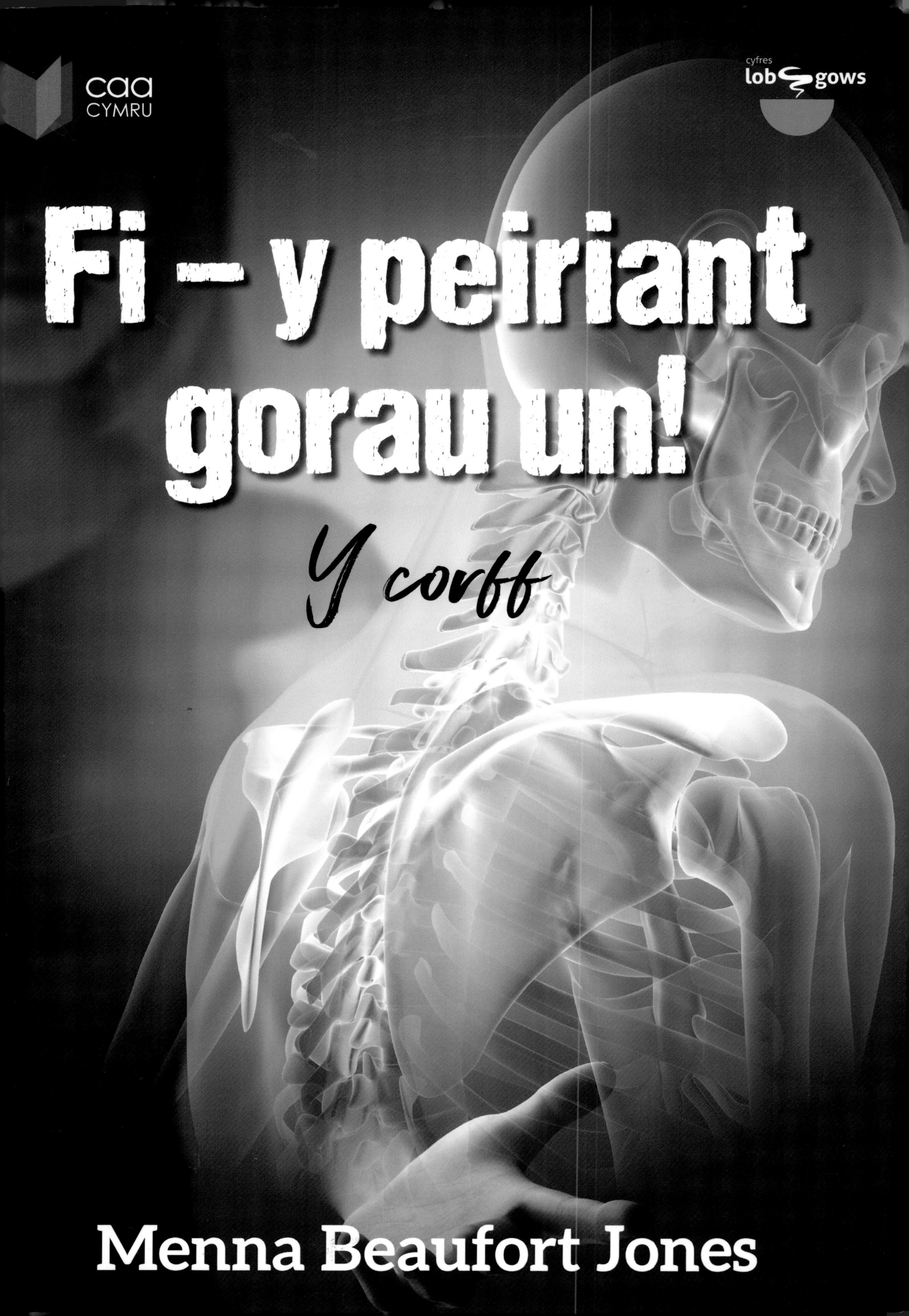

Fi – y peiriant gorau un!

Y corff

Menna Beaufort Jones

Cyhoeddwyd gan CAA Cymru, Prifysgol Aberystwyth, Plas Gogerddan, Aberystwyth SY23 3EB (www.aber.ac.uk/caa).

Ariennir gan Lywodraeth Cymru fel rhan o'i rhaglen gomisiynu adnoddau addysgu a dysgu Cymraeg a dwyieithog.

ISBN: 978-1-84521-684-9

Golygwyd gan Fflur Aneira Davies
Dyluniwyd gan Richard Huw Pritchard
Argraffwyd gan Argraffwyr Cambria

Cydnabyddiaethau

Diolch i'r canlynol am ganiatâd i atgynhyrchu deunyddiau yn y gyfrol hon:

© Comisiynydd Plant Cymru - tud. 34, 36

Gwnaethpwyd pob ymdrech i olrhain a chydnabod deiliaid hawlfraint. Bydd y cyhoeddwr yn falch o wneud trefniadau addas gydag unrhyw ddeiliaid na lwyddwyd i gysylltu â nhw.

Diolch i Gavin Ashcroft, Ceris Davies, Llinos Jones a Mali Williams am eu harweiniad gwerthfawr.